MODERN CONCEPTS IN CHEMISTRY

EDITORS

Bryce Crawford, Jr., University of Minnesota
W. D. McElroy, Johns Hopkins University
Charles C. Price, University of Pennsylvania

LARS MELANDER received his doctorate from the University of Stockholm where he is presently Docent (Lecturer). Acting Head of the Nobel Institute of Chemistry, he was formerly in charge of the nuclear chemistry department of the Nobel Institute of Physics, to which Institute he is still a consultant. In 1953 he was Peter C. Reilly Lecturer in chemistry at the University of Notre Dame. Dr. Melander is the author of a number of papers in nuclear chemistry and physical organic chemistry.

ISOTOPE EFFECTS
ON REACTION RATES

LARS MELANDER

NOBEL INSTITUTE OF CHEMISTRY

STOCKHOLM

THE RONALD PRESS COMPANY · NEW YORK

Library of Congress Catalog Card Number: 60–9663

PRINTED IN THE UNITED STATES OF AMERICA

Preface

The present book, a volume in the series "Modern Concepts in Chemistry," is a brief presentation of the main principles of kinetic isotope effects, with no attempt being made to cover the entire field. The choice of material for a volume like the present one is certainly debatable and is bound to reflect the limitations of the writer's personal choices and interests.

Research work has been referred to primarily to illustrate different aspects of the subject and to show some of the achievements of isotope effects as a scientific tool. It has been considered better to discuss thoroughly a few illuminating procedures and to compare different theoretical methods than to try to compress as many experimental facts as possible into a limited space. For this reason, also, no attempt has been made to give an account of the development of the subject historically. The fact that many experimental results have been left out of the present treatment by no means implies a rejection of them.

Owing to the limited space and the purpose of this volume, much has to be assumed to be familiar to the reader, such as the theory of absolute reaction rates and the kinetic treatment of simple systems of consecutive reactions. The technical experimental aspects of the subject have also been left out in so far as they do not have a direct bearing on the major principles. All these neighboring fields are covered by their own literature, well known to most scientists. Some references to sources of this kind are given in the Introduction.

As to the choice of symbols and other questions of usage, much could be discussed. At present it seems virtually impossible to find a common usage which is in agreement with most published work in the field and in neighboring branches of science. The choice made by the author may not be the best one, but it is a usable compromise. Thus, for instance, in taking the ratio between the specific rates of two isotopic substances, many research workers place the rate of the lighter compound in the numerator. The writer, certainly influenced by his own initial choice, has employed the opposite. The main advantage of the latter is that several related ratios (for instance, k_D/k_H and k_T/k_H) will have the same denominator; i.e., the light isotope, which is generally the main one, is always taken as a basis. One disadvantage is that in the general ratio k_1/k_2, the lower index will correspond to the higher mass number and vice versa. Another is that a "large" isotope effect will correspond to a small ratio. To avoid this, the terms "strong" and "weak" have been used in this sense instead of "large" and "small."

My thanks are due to my wife, who prepared the drawings for the present volume and carried through the numerical computations underlying the diagrams.

Lars Melander

Stockholm, Sweden
 July, 1959

Contents

v

ISOTOPE EFFECTS
ON REACTION RATES

1

Introduction and General References

KINETIC ISOTOPE EFFECTS IN THE STUDY OF CHEMICAL REACTIONS

The first general method for the study of reaction mechanisms was presented by reaction kinetics. Through a study of the rate of production of a certain molecule as a function of the concentrations of the reactants, it was frequently possible to find out how many entities of each kind have to come together. The temperature coefficient of the rate gave information about the energy threshold, which must be surmounted before a proper cluster of reactants may pass into the products. From the general kinetic theory of gases and liquids, the foundation of these theories, it was also possible to conclude that in many complex reactions a collision between the proper individuals is not enough to let them pass into the products, even if sufficient energy is available. Here enters the demand for a certain conformation of the transition state.

Today the term "mechanism" of a chemical reaction generally implies a definite knowledge of the role of each atom of the molecules concerned and also of the electrons, although it must not be forgotten that the latter lack individuality because the uncertainty principle allows much less sharp predictions to be made for an electron than for an atomic nucleus. Generally,

3

we are satisfied at present if we have an approximate idea of the structure and conformation of the transition state, i.e., of the relative positions of the atomic nuclei in the grouping of highest potential energy belonging to the energetically easiest path from reactants to products.

By means of tracer methods it is possible to learn the fate of different atoms. Common knowledge and experience will then tell the investigator which possibilities exist for the transition state. What we know from conformational theory will generally permit approximate predictions of properties for the transition state, since the rearrangements will generally embrace only a minor part of complex molecules.

Different predicted transition states should be tested experimentally. In principle, by virtue of the theory of absolute reaction rates, a known transition state allows the prediction of the absolute reaction rate from its physical parameters and those of the reactants. In order for such a computation to give results accurate enough for a significant comparison with experimental data, the physical parameters have to be known with very great accuracy, and particularly environmental influences in condensed phases will defy purely theoretical treatment.

A much easier way to test a given model is to make relative measurements, and to study the influence of certain internal or external conditions on the relative rate. Many unknown or poorly known magnitudes will be cancelled in the ratio of rates. The external conditions which could be changed are, for instance, the polarity and the ionic strength of the medium; the resultant relative change of the rate will then give information about the charge distribution in the transition state.

An internal change which has been used extensively is the introduction of substituents. This substitution must, of course, be made in some position outside the reaction center in order to maintain the type of reaction. A drawback of the method is that homologues are sometimes too different to allow a dependable comparison; in other words, the cancellation of unknown parts is not good enough. Isotopic substitution, on the other hand, is a much more subtle change. Accordingly, isotopic substitution outside the reaction center will have little influence on the quantitative behavior, except in some cases of "secondary isotope effects." The superiority of the isotope method lies mainly in the possibility of making the substitution within the very reacting center, i.e., at the ends of bonds being broken or created in the reaction. For elements with low atomic weight the effects on reaction rate will then generally be appreciable with a minimum of change in parameters difficult to check, e.g., activity coefficients.

Different models of the transition state will frequently give quite different predictions of such kinetic isotope effects, thus making the measurement of the latter a powerful diagnostic means. It might, however, be useful in this very general introduction to face an inherent limitation of the method, which it has in common with all measurements, absolute or relative, of reaction rates. Since our most useful and probably fairly true theories of reaction rate will bring the latter back onto a thermodynamical equilibrium between the transition state and the reactants and will introduce the time in a very general and formal manner, any rate measurements will probably tell us nothing but the free energy of the transition state. In a more complex reaction the transition state in question is the one corresponding to the rate-determining step, and it is inherently impossible to draw any conclusions about the mechanism of its formation, for instance, via prior, fast reaction steps. It means also that we cannot always obtain information about everything which happens even in the rate-determining step itself. Thus the breaking of a bond may occasionally not be appreciably advanced in the transition state but nevertheless may be completed in the same reaction step. The potential-energy maximum might be reached early as a result of other energy contributions. We can never hope to verify such a bond breaking in the rate-determining step by the observation of an appreciable isotope effect. Overlooking these principles may lead to inadequate conclusions.

GENERAL REFERENCES

A very recent survey of the subject covered by the present volume has been given by Bigeleisen and Wolfsberg (14). This survey should be consulted particularly for details and recent development of the theory as designed by those writers and for the theoretical treatment of unimolecular reactions at low pressures, which has been omitted here.

Useful keys to work on isotope effects are a review by Wiberg (101) which deals with hydrogen isotope effects, and the monograph by Roginsky (80), which has been translated into English.

The existing theoretical treatment of isotope effects is founded on the theory of absolute reaction rates, as presented in the classical treatment, *The Theory of Rate Processes*, by Glasstone, Laidler, and Eyring (39).

For the kinetics of complex reactions and other problems of reaction kinetics the reader is referred to the well-known monograph by Frost and Pearson (37).

In the theoretical computation of isotope effects the frequencies of the various modes of vibration of molecules are required. As important sources may be mentioned Herzberg's *Infrared and Raman Spectra of Polyatomic Molecules* (48) and the monographs on infrared spectroscopy by Bellamy (3) and Brügel (20). A very useful paper by Urey (96) contains vibrational data and partition function ratios for many of the simpler isotopic molecules.

The geometrical dimensions of many organic molecules might be obtained from the Appendix of Wheland's *Resonance in Organic Chemistry* (100).

Except for general chemical techniques, the experimental work on isotope effects requires synthetic work with and quantitative determination of nuclides. The recent comprehensive *Organic Syntheses with Isotopes* by Murray and Williams (70) will be helpful in the first respect. It also contains an introduction with several useful pieces of practical information, for instance, numerous references to assay methods for stable as well as radioactive nuclides.

2

Prediction of Rate-Constant Ratios from Molecular Data

Limitations. It is evident that any calculation of the influence of isotopic mass on the rate of a reaction has to make use of a general theory of reaction rate. It is also easy to see that a theory is required which makes use of a very detailed description of the roles of the individual atoms and the forces exerted by the electron cloud, acting like an elastic glue keeping the atoms of a molecule together. The absolute reaction rate theory, as presented, for instance, by Glasstone, Laidler, and Eyring (39), has proved very useful for this purpose. Its limitation in practical cases lies in our incomplete knowledge of the actual transition state and the impossibility at present of making a quantum-mechanical calculation accurate enough for any but the simplest systems. Contrary to the case with several complex but stable molecules, we still lack information about the conformation and even the structure of most transition states. In order to predict the reaction rate and the isotope effect of a certain reaction accurately, detailed information of the transition state is required. On the other hand, and this is at present most important, experimental results will serve as a check of a tentative transition state and help us to select the true one from the few which seem likely. A qualitative agreement is then generally sufficient, and more can generally not be expected, since several of the physical data of a tentative transition state will have to be roughly estimated.

7

Our estimates will generally be based on what we know about stable molecules. Most bonds which are not directly involved in the reaction will behave quite normally as to length and force constant. For bonds which are directly involved in the reaction, the length will generally be deemed "increased" and the force constants "weakened" if the corresponding mode of vibration has not disappeared completely as a consequence of the transformation of the corresponding vibrational coordinate into the coordinate of decomposition. Thus general and qualitative statements stand for exact figures, and frequently the only possibility is to make calculations on the limiting assumption that zero frequency results from the weakening. Since the results are often much less sensitive to moderate changes in the molecular dimensions, a fair estimate of bond lengths might suffice if required, and sometimes the change might be completely neglected.

Two more factors, which cannot be calculated without a thorough knowledge of the potential-energy surface, are the transmission coefficient and the amount of tunneling through the potential-energy barrier. In all ordinary cases very little can be done to account for these effects, but some theoretical evidence and general experience indicate that the transmission coefficient is not very sensitive to isotopic mass in cases of experimental interest and will consequently be cancelled in the isotope effect ratio, and the tunneling contribution is generally negligible in the absolute rate.

Invariance of Potential-Energy Surface. Before proceeding further, a very important general principle concerning isotopic molecules should be stated. A difference in the mass of an atomic nucleus will have very little influence on the structure of the electron cloud surrounding the nucleus. Thus, for instance, in the theory of the electronic structure of the hydrogen atom, the mass of the nucleus enters the calculations only via the reduced mass of the system of nucleus plus electron. If the reduced mass is denoted by μ, we have

$$\frac{1}{\mu} = \frac{1}{m_{\text{electron}}} + \frac{1}{m_{\text{nucleus}}}$$

Since for any isotope of hydrogen the last term is about $1/2000$ or less of the first one to the right, it is obvious that a variation in the mass number of the hydrogen atom between one and infinity will cause a very slight variation in μ.

On the other hand, if all the atom is accelerated as a unit, the sum $(m_{\text{electron}} + m_{\text{nucleus}})$ and hence m_{nucleus} almost alone will be decisive.

If these principles are carried over to molecules, the electronic structure and hence the forces which hold the atoms together will be nearly independent of changes in the masses of the atomic nuclei caused by isotopic substitution. This means that the potential-energy surface and hence the interatomic distances and the vibrational force constants could be treated as invariant under isotopic substitution with an accuracy which is sufficient for our purposes.* In the case of vibrations, when the atomic nuclei are in accelerated movement, however, the inert masses will have a strong influence.

It is obvious from the preceding statements that there exists a relation between the moments of inertia of isotopic molecules, since the geometrical arrangement of the atomic nuclei will be the same, and there will also be a relation between the vibrational frequencies owing to the invariant force constants. Use will be made of such relations in the following.

PRIMARY EXPRESSION

Absolute Reaction Rate. The complete rate expression for the reaction

$$A + B + \cdots \rightarrow \text{products}$$

---

* Owing to vibrational anharmonicity, however, a potential-energy minimum does not define the (average) interatomic distance perfectly. Thus Halevi (43) claims that the difference in carbon-protium and carbon-deuterium distance might be large enough (even in the vibrational ground state) to give rise to measurable chemical effects; cf. p. 96. Hitherto such effects have generally been completely ignored and will be so in this volume unless otherwise stated.

in which the transition state $M^{\ddagger}$ is formed according to the equilibrium

$$A + B + \cdots \;\; \rightleftarrows \;\; M^{\ddagger} + N + \cdots$$

is given in the following way by the theory of absolute reaction rates (39):

$$\text{Rate} = \left[1 - \frac{1}{24}\left(\frac{h\nu_L^{\ddagger}}{kT}\right)^2\right] \kappa \frac{kT}{h} K^{\ddagger} \frac{[A][B]\cdots}{[N]\cdots} \times \frac{\alpha_A\alpha_B\cdots}{\alpha^{\ddagger}\alpha_N\cdots}$$

$$(2\text{-}1)$$

The expression in square brackets is a correction factor for tunneling or leakage through the energy barrier. $\nu_L^{\ddagger}$ is the (imaginary) frequency of vibration along the decomposition coordinate, h is Planck's constant, k is Boltzmann's constant, and T is the absolute temperature. κ is the transmission coefficient, or the fraction of transition states passing the energy barrier in the forward direction which will lead to completed reaction. $K^{\ddagger}$ is a kind of thermodynamical equilibrium constant of the activation equilibrium (see below). The small square brackets symbolize concentration. α denotes activity coefficient, that of the transition state being written $\alpha^{\ddagger}$.

All pertinent information about the strict derivation of Eq. (2-1) will be found in the book by Glasstone, Laidler, and Eyring (39), and only the following comments are made in this connection.

In order to impart an intuitive feeling of the principles underlying Eq. (2-1) the following reasoning might be allowed. The concentration of the transition state is determined by the equilibrium

$$\frac{[M^{\ddagger}][N]\cdots}{[A][B]\cdots} \times \frac{\alpha^{\ddagger}\alpha_N\cdots}{\alpha_A\alpha_B\cdots} = K^{\ddagger}\frac{kT}{h\nu_L^{\ddagger}} \qquad (2\text{-}2)$$

in which expression $\nu_L^{\ddagger}$ is best thought of as a very low but real frequency. The rate of passage over the energy barrier (in concentration units) will be $\nu_L^{\ddagger}[M^{\ddagger}]$, and the fraction κ of this rate is effective. Quantum-mechanical tunneling through the barrier, finally, will slightly increase the reaction rate.

It is evident from expression (2–1) that it is primarily the position of the thermodynamical equilibrium that determines the reaction rate and its changes with changes in the environment. In the same way the influence of isotopic substitution on reaction rate will be exercised via its influence on $K^\ddagger$. The following calculations will deal mainly with $K^\ddagger$, which might be computed from statistical mechanics. First, however, something must be said about tunneling and the transmission coefficient.

Tunneling. In most ordinary chemical reactions, tunneling seems to be without importance. The imaginary frequency $\nu_L{}^\ddagger$ of vibration along the coordinate of decomposition corresponds to the curvature of the potential-energy surface along the same coordinate. Since this curvature is concave downwards, the frequency is imaginary and has the same absolute value as if the same curvature were concave upwards, giving rise to an ordinary vibration with a real frequency.

In order to obtain an idea of the curvature required to give a serious leakage effect, let us consider the frequency which makes $\boldsymbol{h} \left| \nu_L{}^\ddagger \right| = \boldsymbol{k}T$. Such a $\nu_L{}^\ddagger$ will obviously give about 4 per cent tunneling. $\left| \nu_L{}^\ddagger \right|$ becomes equal to about 200 cm^{-1} at room temperature. This is a fairly low frequency for vibrations within molecules, but the potential-energy surface will generally be fairly flat in the direction of the reaction coordinate at the saddle point corresponding to the transition state.

In view of the general experience that tunneling seems to be without importance in the majority of well-investigated chemical reactions, it will not be discussed further, since it is then also without importance for the isotope effect, which depends only on the ratio between two correction factors of the type shown in Eq. (2–1). For two isotopic reactions, the $\nu_L{}^\ddagger$ values are in the inverse ratio of the square roots of the reduced masses along the decomposition coordinate. An accurate calculation of the effect requires a thorough knowledge of the shape of the potential-energy surface and consequently is not possible in most cases.

Transmission Coefficient. The transmission coefficient is perhaps the weakest point in the present theory. For an accurate calculation the potential-energy surface has to be known, and a rigorous calculation is difficult even then. Hirschfelder and Wigner (50) have considered the problem and concluded that at not too low temperatures isotopic deviations should not be serious, provided there is a thermal distribution of energy as in practical cases. In the following we simply assume that the transmission coefficients introduce no isotope effect.

Activation Equilibrium. We have now localized the influence of isotopic mass on reaction velocity to the constant $K^{\ddagger}$ and the activity coefficients. In the beginning we shall assume that ideal gaseous conditions prevail, and that the activity coefficients are all equal to unity. The ratio between the specific rates of two isotopic reactions will then equal the ratio between the two constants $K^{\ddagger}$.

For an ordinary equilibrium, the concentration * equilibrium constant might be computed from the partition functions Q° for unit volume:

$$K_c = \frac{\Pi Q^{\circ}_{\text{products}}}{\Pi Q^{\circ}_{\text{reactants}}} e^{-E_{\text{class}}/RT} \qquad (2\text{--}3)$$

In this case the energy levels of the products and those of the reactants are assumed to be computed with reference to the energy origin of the respective classical oscillators; i.e., the bottom of the potential-energy curve. The exponential accounts for the corresponding difference in energy origin between the products and the reactants, E_{class} being the classical heat of reaction at the absolute zero (in molar amount) and R the gas constant. From the previous discussion it is evident that E_{class} can be treated as independent of the isotopic masses.

A similar computation is possible, at least in principle, for the activation equilibrium. Owing to the specific role of the transition state in a chemical reaction, however, the partition

---

* The proper concentration unit to be used in this connection is 1 molecule per unit volume.

function of that state has to be treated with slight caution. Contrary to ordinary molecules, which correspond to true minima in the potential-energy surface, the transition state corresponds to a saddle point; i.e., it is unstable toward movements along the decomposition coordinate mentioned above. If the transition state is compared to an ordinary molecule, the movement in question could be compared to a kind of vibrational movement which tears the molecule apart. This particular vibration, which has an imaginary frequency $\nu_L^\ddagger$, is best treated as a classical vibration, for which the partition function would be $kT/h\nu_L^\ddagger$. In order to find the rate of passage over the barrier we have to multiply by $\nu_L^\ddagger$. The product $\nu_L^\ddagger \times kT/h\nu_L^\ddagger$ gives rise to the universal constant kT/h in Eq. (2–1).

When the activation equilibrium constant is discussed, the constant $K^\ddagger$ is generally referred to. This constant is what remains after the above partition function for the movement along the decomposition coordinate has been removed from the complete equilibrium constant. The partition function of the transition state used in Eq. (2–3) for the calculation of $K^\ddagger$ should consequently contain no further contribution from this degree of freedom. Otherwise the transition state could be treated as an ordinary molecule. Its dimensions and vibrational frequencies have to be found out from the potential-energy surface or estimated from analogies with ordinary, stable molecules.

The partition function for unit volume of an ordinary molecule has the following general appearance to a sufficient degree of approximation:

$$Q° = Q_{\text{trans}} Q_{\text{rot}} Q_{\text{vibr}} \times \frac{1}{V} =$$

$$= \frac{(2\pi MkT)^{3/2}}{h^3} \times \frac{g_{\text{el}} \, g_{\text{nuc}}}{s} \times \frac{8\pi^2(8\pi^3 ABC)^{1/2}(kT)^{3/2}}{h^3} \times$$

$$\times \prod_i e^{-\frac{1}{2}h\nu_i/kT}\left(1 - e^{-h\nu_i/kT}\right)^{-1} \qquad (2\text{–}4)$$

The first fraction on the second line of Eq. (2–4) is the translational partition function divided by the volume, Q_{trans}/V, and contains the molecular mass M. k, T, and h have their usual meaning as above. g_{el} and g_{nuc} account for the electronic statistical weight of the ground state and degeneracy due to different orientations of the nuclear spins, respectively. At ordinary temperatures, most molecules have such an extremely low population in excited electronic states that the electronic partition function is simply g_{el}. s is the symmetry number, or the number of indistinguishable positions the molecule might take. A, B, and C are the principal moments of inertia. Modifications of the third fraction on the second line have to be made for molecules with less than three degrees of rotational freedom. A monatomic molecule, for instance, will have no such degree of freedom, and a diatomic or other linear molecule will have two. For hydrogen molecules the rotational partition function deviates from its classical value even at temperatures as high as room temperature. For approximate calculations, however, these deviations need not be taken into consideration if the temperature is not considerably below room temperature. The product in Eq. (2–4) contains factors corresponding to all vibrational degrees of freedom. As the expression is written, the energy origin is that of a classical oscillator, i.e., the bottom of the potential-energy curve. The lowest vibrational level of the molecule will consequently be assigned an energy value of $\sum_i \frac{1}{2} h \nu_i$.

If expression (2–4) is to be used for a transition state, the only modification will be that the number of ν_i's will be one less than that for an ordinary molecule, as explained above.

Rate-Constant Ratio. We are now going to use Eq. (2–4) in setting up relation (2–3) for the equilibrium between the transition state and the reactants in two isotopic reactions:

$$A_{(1)} + B + \cdots \xrightarrow{k_1} (\text{products})_1$$
$$A_{(2)} + B + \cdots \xrightarrow{k_2} (\text{products})_2$$

where the specific rates are given above the arrows. The transition-state equilibria are assumed to be, respectively,

$$A_{(1)} + B + \cdots \underset{\longleftarrow}{\overset{K_1^\ddagger}{\rightleftarrows}} M_{(1)}^\ddagger + N + \cdots$$

$$A_{(2)} + B + \cdots \underset{\longleftarrow}{\overset{K_2^\ddagger}{\rightleftarrows}} M_{(2)}^\ddagger + N + \cdots$$

(As explained above, $K^\ddagger$ is nearly, but not exactly, an ordinary equilibrium constant.) The indices 1 and 2 denote two species which are isotopic. Hence it is assumed that of the reactants only A occurs in two different isotopic forms, and that possible by-products in the formation of the transition state do not differ in isotopic composition. These assumptions are very likely to be applicable to most situations in practice.

Neglecting any isotope effect arising from tunneling, the transmission coefficients or the activity coefficients, we can write

$$\frac{k_1}{k_2} = \frac{K_1^\ddagger}{K_2^\ddagger} = \frac{Q_1^{o\ddagger}}{Q_2^{o\ddagger}} \times \frac{Q_{A(2)}^o}{Q_{A(1)}^o} =$$

$$= \left(\frac{M_1^\ddagger}{M_2^\ddagger}\right)^{3/2} \frac{s_2^\ddagger}{s_1^\ddagger} \left(\frac{A_1^\ddagger B_1^\ddagger C_1^\ddagger}{A_2^\ddagger B_2^\ddagger C_2^\ddagger}\right)^{1/2} \prod_i^{3n\ddagger-7} \left(\frac{e^{-\frac{1}{2}u_{i(1)}^\ddagger}}{e^{-\frac{1}{2}u_{i(2)}^\ddagger}} \times \frac{1 - e^{-u_{i(2)}^\ddagger}}{1 - e^{-u_{i(1)}^\ddagger}}\right) \times$$

$$\times \left(\frac{M_2}{M_1}\right)^{3/2} \frac{s_1}{s_2} \left(\frac{A_2 B_2 C_2}{A_1 B_1 C_1}\right)^{1/2} \prod_i^{3n-6} \left(\frac{e^{-\frac{1}{2}u_{i(2)}}}{e^{-\frac{1}{2}u_{i(1)}}} \times \frac{1 - e^{-u_{i(1)}}}{1 - e^{-u_{i(2)}}}\right) =$$

$$= \frac{s_2^\ddagger}{s_1^\ddagger} \times \frac{s_1}{s_2} \left(\frac{M_1^\ddagger}{M_2^\ddagger} \times \frac{M_2}{M_1}\right)^{3/2} \left(\frac{A_1^\ddagger B_1^\ddagger C_1^\ddagger}{A_2^\ddagger B_2^\ddagger C_2^\ddagger} \times \frac{A_2 B_2 C_2}{A_1 B_1 C_1}\right)^{1/2} \times$$

$$\times \prod_i^{3n\ddagger-7} \frac{\sinh \frac{1}{2}u_{i(2)}^\ddagger}{\sinh \frac{1}{2}u_{i(1)}^\ddagger} \prod_i^{3n-6} \frac{\sinh \frac{1}{2}u_{i(1)}}{\sinh \frac{1}{2}u_{i(2)}} \tag{2-5}$$

This relation we shall call the primary one. The symbols are analogous to those in Eq. (2–4), although $h\nu/kT$ has been replaced by the symbol u for the sake of brevity. As in any type of equilibrium, the factors g_{nuc} cancel each other within each $K^\ddagger$, and the g_{el} factors are cancelled between the two transition states and between the two molecules of type A, respectively, because isotopic species have the same electronic structure. If the transition state or the molecule A has less than three degrees of rotational freedom, the factor including the moments of inertia and the number of modes of vibration will have to be changed correspondingly. The number of atoms in the transition state has been denoted by $n^{\ddagger}$ and that in the reactant A by n. For a transition state, which is not linear, the number of vibrational modes to account for explicitly is one less than for an ordinary molecule, as explained above; hence the number $3n^{\ddagger} - 7$.

From this point there will be some branching in our discussion as required by the nature of the problems to which the theory is to be applied and by the degree of accuracy desired.

Some General Approximations. In general, little is known of the vibrational frequencies of the isotopic reactants and still less of those of the transition state. A common approximation in this situation is to assume that all but one of the vibrational contributions disappear by cancellation. That one corresponds to the stretching mode of the bond of the reactant which is going to be cleaved in the reaction. This movement corresponds to the aperiodic motion of the transition state along the reaction coordinate, which does not show up in Eq. (2–5), and hence there can be no cancellation. All other frequencies are assumed to be cancelled, either between one of the reactants and the corresponding transition state or between two isotopic species of the same kind. The latter possibility is assumed to be realized for those parts of the transition state which do not originate from the isotopic reactant and which make $n^{\ddagger} > n$, for instance in a bimolecular reaction. Such vibrations of the transition state are likely to be approximately non-isotopic. (Of course, a rough approximation is always involved when "bond vibrations" or vibrations of minor parts of a molecule are treated as if they were the true normal modes of vibration.) The resulting expression is

$$\frac{k_1}{k_2} \times \frac{s_2}{s_1} \times \frac{s_1^{\ddagger}}{s_2^{\ddagger}} =$$
$$= \left(\frac{M_1^{\ddagger}}{M_2^{\ddagger}} \times \frac{M_2}{M_1}\right)^{3/2} \left(\frac{A_1^{\ddagger}B_1^{\ddagger}C_1^{\ddagger}}{A_2^{\ddagger}B_2^{\ddagger}C_2^{\ddagger}} \times \frac{A_2B_2C_2}{A_1B_1C_1}\right)^{1/2} \frac{\sinh\frac{1}{2}u_{k(1)}}{\sinh\frac{1}{2}u_{k(2)}} \quad (2\text{–}6)$$

where $u_{k(1)}$ and $u_{k(2)}$ correspond to the stretching frequencies in question. If the frequency is known only for one of the two isotopic reactants, the other is approximately calculated from this value as if the reactant were a diatomic molecule consisting merely of the atoms (or sometimes the fragments) at each end of the pertinent bond.

Another approximation involves the assumption that the molecular masses and the moments of inertia give no contribution to the isotope effect, i.e., that the two parentheses of Eq. (2–6) could be replaced by unity. In unimolecular decompositions, for instance, the first parenthesis equals unity exactly, and the second does so at least approximately. The remaining expression is

$$\frac{k_1}{k_2} \times \frac{s_2}{s_1} \times \frac{s_1^{\ddagger}}{s_2^{\ddagger}} = \frac{\sinh \frac{1}{2} u_{k(1)}}{\sinh \frac{1}{2} u_{k(2)}} = \frac{\sinh \dfrac{hc\bar{\nu}_{k(1)}}{2kT}}{\sinh \dfrac{hc\bar{\nu}_{k(2)}}{2kT}} \qquad (2\text{–}7)$$

which has been applied to hydrogen and carbon isotope effects by Eyring and Cagle (35). (c is the velocity of light and $\bar{\nu}$ the wave number.)

ISOTOPIC HYDROGEN IN HEAVY MOLECULES

The primary hydrogen isotope effect in reactions of comparatively heavy molecules has been extensively studied and deserves particular attention here. It might be useful to rewrite the primary expression (2–5) in the following way:

$$\frac{k_1}{k_2} \times \frac{s_2}{s_1} \times \frac{s_1^{\ddagger}}{s_2^{\ddagger}} = \left(\frac{M_1^{\ddagger}}{M_2^{\ddagger}} \times \frac{M_2}{M_1}\right)^{3/2} \left(\frac{A_1^{\ddagger}B_1^{\ddagger}C_1^{\ddagger}}{A_2^{\ddagger}B_2^{\ddagger}C_2^{\ddagger}} \times \frac{A_2B_2C_2}{A_1B_1C_1}\right)^{1/2} \times$$

$$\times \prod_i^{3n^{\ddagger}-7} \frac{1 - e^{-u_{i(2)}^{\ddagger}}}{1 - e^{-u_{i(1)}^{\ddagger}}} \prod_i^{3n-6} \frac{1 - e^{-u_{i(1)}}}{1 - e^{-u_{i(2)}}} \times$$

$$\times \exp \left\{ -\frac{1}{2} \left[\sum_i^{3n^{\ddagger}-7} \left(u_{i(1)}^{\ddagger} - u_{i(2)}^{\ddagger}\right) - \sum_i^{3n-6} \left(u_{i(1)} - u_{i(2)}\right) \right] \right\} \qquad (2\text{–}8)$$

The symmetry numbers are fairly trivial and will cause no isotope effect. For instance, if the reaction velocity "per position" in a molecule is considered, the correction for differences in symmetry between isotopic species will generally be taken into account automatically. It might, however, sometimes be safe to go back to the rigid statistical treatment.

The last exponential in Eq. (2–8) contains in its exponent the difference between the transition state and the reactant of the differences between the zero-point energies of the isotopic cases 1 and 2, all divided by kT. In Fig. 2–1 the curve shows

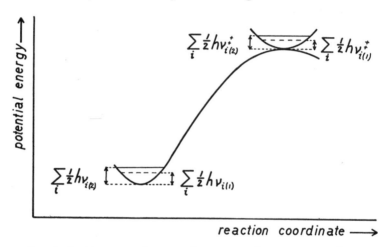

Fig. 2–1. Diagram showing the potential energy as a function of the reaction coordinate. Zero-point-energy levels indicated.

the potential energy of a reacting system as a function of the reaction coordinate. The curve could also be said to show the profile of the potential-energy surface along the reaction path. The zero-point levels of two isotopic species have been represented for the initial state as well as for the transition state. It should be remembered that the latter corresponds to a saddle point of the surface and has vibrational modes in directions other than the decomposition coordinate, as has been symbolically indicated. A remarkable thing is that the activation energy, i.e., the difference in energy level between the initial and transition states, does not enter our expression and has no direct bearing on the magnitude of the isotope effect.

Approximations. For fairly heavy molecules the substitution of one or a few hydrogen atoms for heavy isotopes will materially affect neither the molecular mass nor the moments of inertia. Furthermore, for both magnitudes there will be extensive cancellation between the ratio for the transition state and that for the reactant, provided these two entities are not too dissimilar. Thus the two parentheses of Eq. (2–8) containing the molecular masses and the moments of inertia will be approximately equal to unity.

For a discussion of the vibrational products of Eq. (2–8) it is important to remember that the spacing between the vibrational levels of ordinary molecules will generally be considerably larger than the quantity kT at room temperature. This means that most oscillators are in their zero-point levels. The exponential e^{-u} will be small compared to unity. Molecular vibrations will quite generally have frequencies above 400 cm^{-1}, and those which involve hydrogen to any appreciable extent, i.e., those which are sensitive to the mass of hydrogen, will have considerably higher frequencies, for instance 700 cm^{-1}. For these two frequencies the exponential e^{-u} has the values 0.15 and 0.03, respectively, at 300°K. It is obvious that the vibrational product of the reactant will be close to unity and fairly insensitive to the isotopic mass of the hydrogen at temperatures not too elevated.

For the transition state the same holds primarily, in principle, although it must be remembered that some vibrational modes are likely to behave in a manner different from that in ordinary molecules. The vibration corresponding to movement along the decomposition coordinate was excluded and thus does not matter, but there are, for instance, also bending vibrations of bonds in the state of being ruptured. No definite predictions can be made without a detailed knowledge of the potential-energy surface. Frequently it is assumed that the last-mentioned bending modes behave mainly in the same way as they do in the reactant. The products in Eq. (2–8) could under such circumstances be equated to unity.

Zero-Point-Energy Effect. Under the presumptions above we are left with the zero-point exponential, which is by far the most important factor in discussions of primary hydrogen isotope effects, i.e., isotope effects of reactions in which hydrogen is directly involved.

$$\frac{k_1}{k_2} \times \frac{s_2}{s_1} \times \frac{s_1^{\ddagger}}{s_2^{\ddagger}} =$$

$$= \exp\left\{ -\frac{1}{2}\left[\sum_i^{3n^{\ddagger}-7}(u_{i(1)}^{\ddagger} - u_{i(2)}^{\ddagger}) - \sum_i^{3n-6}(u_{i(1)} - u_{i(2)}) \right] \right\} \quad (2\text{–}9)$$

The characteristic feature of the exponential is most easily seen if the isotopic reactant also serves as a model for the transition state, as in a unimolecular decomposition, and, approximately, if the involvement of other molecules in the transition state of any reaction is so weak that the transition state might be regarded as one decomposing molecule plus essentially free other molecules. The properties of the latter (by our assumption not isotopic) molecules will, of course, immediately be cancelled as they occur analogously in both transition states. Under these presumptions the numbers $n^{\ddagger}$ and n might be considered equal, and consequently there is one half-quantum less in the zero-point energy of the transition state than in the reactant. If otherwise the frequencies in the latter match those of the former, we have the expression

$$\frac{k_1}{k_2} \times \frac{s_2}{s_1} \times \frac{s_1^{\ddagger}}{s_2^{\ddagger}} = \exp\left\{ \frac{1}{2}(u_{k(1)} - u_{k(2)}) \right\} =$$

$$= \exp\left\{ \frac{hc}{2kT}(\bar{\nu}_{k(1)} - \bar{\nu}_{k(2)}) \right\} \quad (2\text{–}10)$$

where c is the velocity of light and $\bar{\nu}$ denotes wave number. $\bar{\nu}_k$ is generally the stretching frequency of a bond linking hydrogen to some other atom, and this bond is cleaved in the reaction. The corresponding stretching mode in the transition state is the aperiodic motion along the reaction coordinate, which should be excluded from the transition state sum of Eq. (2–9).

It is evident that Eyring-Cagle's expression (2–7) is equivalent to (2–10) if the frequency is high enough, because at high x we have $\sinh x \approx \frac{1}{2}e^x$.

If the stretching frequencies of both isotopic bonds are not known, one might be calculated from the other by means of the reduced masses of the systems hydrogen plus heavy residue. For the reason in the case discussed on p. 9 the small mass will be decisive, and for crude approximations the hydrogen mass is used for the reduced mass. Thus

$$\frac{u_{k(1)}}{u_{k(2)}} = \frac{\nu_{k(1)}}{\nu_{k(2)}} = \left(\frac{m_{H(2)}}{m_{H(1)}}\right)^{\frac{1}{2}} \tag{2–11}$$

where m_H denotes the hydrogen mass.

As pointed out above, the bending vibrations of the cleaving bond might behave differently. If they have their frequency increased in the transition state, this increase of the zero-point energy on passing into the transition state might more or less cancel the influence of the extra term in the zero-point energy of the reactant, a weakening of the isotope effect being the result. This certainly occurs in many cases. It is of more interest, however, to see how much the isotope effect might be strengthened by the opposite effect.

Effect with Zero Bending Frequencies. If the bending frequencies of the cleaving bond are best approximated by zero, the assumptions leading from Eq. (2–8) to Eq. (2–9) are no longer valid. Two factors of the type $(1 - e^{-u_{(2)}\ddagger})/(1 - e^{-u_{(1)}\ddagger})$ cannot be approximated by unity. Such a factor approaches the value $u_{(2)}\ddagger/u_{(1)}\ddagger$ when the force constant and hence the u's approach zero. Since we have assumed that the hydrogen is attached to a heavy residue, we can approximate this ratio by $(m_{H(1)}/m_{H(2)})^{\frac{1}{2}}$. The two bending modes will thus introduce a factor m_1/m_2, if we drop the letter in the indices. Since, according to our simple model, two vanishing frequencies have been assigned to the transition state, there will now be three unmatched terms in the zero-point energy of the reactant, giving

$$\frac{k_1}{k_2} \times \frac{s_2}{s_1} \times \frac{s_1^{\ddagger}}{s_2^{\ddagger}} = \frac{m_1}{m_2} \exp \left\{ \frac{1}{2} \sum_i^3 (u_{i(1)} - u_{i(2)}) \right\} =$$

$$= \frac{m_1}{m_2} \exp \left\{ \frac{hc}{2kT} \sum_i^3 (\bar{\nu}_{i(1)} - \bar{\nu}_{i(2)}) \right\} \qquad (2\text{--}12)$$

Possibly unknown frequencies, either bending or stretching, could be calculated from known isotopic ones by means of Eq. (2–11).

Approximate Rate-Constant Ratios for Hydrogen.

Equations (2–10) and (2–12) could be used for rough calculations of k_D/k_H and k_T/k_H at different temperatures if some general values could be assigned to the frequencies. For protium bonded to carbon the following values are of some general applicability: stretching, 3000 cm^{-1}; two bending together, 2000 cm^{-1}. The frequencies of the heavy-hydrogen oscillators are obtained by means of Eq. (2–11). The rate-constant ratios for different temperatures are shown in Table 2–1. The difference between

TABLE 2–1

Hydrogen Isotope Effects Calculated by Means of the Approximate
Equations (2–10), (2–11), and (2–12)

°C	k_D/k_H		k_T/k_H	
	Eq. (2–10)	Eq. (2–12)	Eq. (2–10)	Eq. (2–12)
0	0.10	0.042	0.035	0.011
25	0.12	0.058	0.047	0.018
50	0.14	0.077	0.059	0.027
75	0.16	0.097	0.073	0.038
100	0.18	0.12	0.087	0.051
125	0.20	0.14	0.10	0.066
150	0.22	0.17	0.12	0.083

the results according to Eqs. (2–10) and (2–12) is, of course, due to the difference in underlying assumptions. In the former case only the zero-point energy of the stretching vibration is assumed to be lost on passing into the transition state, in the latter case that of all three hydrogen vibrations.

Interrelation of Deuterium and Tritium Isotope Effects. In this connection it might be of interest to consider the possibilities of interrelating different analogous isotope effects, particularly k_D/k_H and k_T/k_H measured for the same reaction at the same temperature.   Swain et al. (95) have recently discussed the problem.

If all vibrations in the reactant and all real ones in the transition state have high frequencies relative to the temperature, the isotope effect is determined by the zero-point-energy exponential as shown in Eq. (2–9).   For hydrogen attached to carbon or some heavier atom, analogous frequencies, bending or stretching, could be calculated by means of Eq. (2–11) thus:

$$u_{(1)} - u_{(2)} = u_{(2)} \left[\left(\frac{m_2}{m_1} \right)^{\frac{1}{2}} - 1 \right]$$

where m denotes hydrogen mass. If the next time we study the isotope effect between species (3) and (2), we could write in the same way

$$u_{(3)} - u_{(2)} = u_{(2)} \left[\left(\frac{m_2}{m_3} \right)^{\frac{1}{2}} - 1 \right]$$

Since the same square bracket factors will occur with each mode of hydrogen vibration, and other frequencies are assumed to be cancelled, the exponentials will be the same for each isotope effect except for these factors in the exponent.   We could consequently write

$$\frac{\log \left(\dfrac{k_1}{k_2} \times \dfrac{s_2}{s_1} \times \dfrac{s_1^{\ddagger}}{s_2^{\ddagger}} \right)}{\log \left(\dfrac{k_3}{k_2} \times \dfrac{s_2}{s_3} \times \dfrac{s_3^{\ddagger}}{s_2^{\ddagger}} \right)} = \frac{\left(\dfrac{m_2}{m_1} \right)^{\frac{1}{2}} - 1}{\left(\dfrac{m_2}{m_3} \right)^{\frac{1}{2}} - 1}$$

If the first ratio, k_1/k_2, compares tritium and protium, and the second, k_3/k_2, compares deuterium and protium, we obtain, with the approximate masses $m_1 = 3$, $m_2 = 1$, and $m_3 = 2$,

$$\frac{\log \left(\dfrac{k_T}{k_H} \times \dfrac{s_H}{s_T} \times \dfrac{s_T^\ddagger}{s_H^\ddagger}\right)}{\log \left(\dfrac{k_D}{k_H} \times \dfrac{s_H}{s_D} \times \dfrac{s_D^\ddagger}{s_H^\ddagger}\right)} = 1.44 \qquad (2\text{--}13)$$

If some frequencies are low, the relation is no longer that simple, but generally the deviations are quite small and Eq. (2–13) is approximately valid. For example, if Eq. (2–12) is a good approximation, we would have instead

$$\frac{\log \left(\dfrac{m_H}{m_T} \times \dfrac{k_T}{k_H} \times \dfrac{s_H}{s_T} \times \dfrac{s_T^\ddagger}{s_H^\ddagger}\right)}{\log \left(\dfrac{m_H}{m_D} \times \dfrac{k_D}{k_H} \times \dfrac{s_H}{s_D} \times \dfrac{s_D^\ddagger}{s_H^\ddagger}\right)} = 1.44$$

or

$$\frac{k_T}{k_H} \times \frac{s_H}{s_T} \times \frac{s_T^\ddagger}{s_H^\ddagger} = \frac{3}{(2)^{1.44}} \left(\frac{k_D}{k_H} \times \frac{s_H}{s_D} \times \frac{s_D^\ddagger}{s_H^\ddagger}\right)^{1.44} =$$

$$= 1.11 \left(\frac{k_D}{k_H} \times \frac{s_H}{s_D} \times \frac{s_D^\ddagger}{s_H^\ddagger}\right)^{1.44}$$

The deviation is not worse than that caused by an 11 per cent error in the ratio k_T/k_H.

Since the two heavy molecules are analogous, we have of course $s_T = s_D$ and $s_T^\ddagger = s_D^\ddagger$.

Hydrogen Isotope Effect with Three-Center Transition State. It might now be of interest to consider the approximate isotope effect of a three-center reaction of related kind. This type arises when some kind of non-isotopic acceptor for the hydrogen is directly involved in the transition state, and it is probably the best picture of various hydrogen-abstraction reactions. The presence of an acceptor, which will generally consist of a carbon or an oxygen atom or a still heavier atom plus possible ligands, will further increase the moments of inertia of the transition state and make the difference in mass of the hydrogen to be transferred still more immaterial for their magnitude. New frequencies will be introduced from the acceptor part. If it is complex it has more or less "internal" vibrations, which will be cancelled between the two isotopic cases, because they are approximately not influenced by the isotopic mass of the hydrogen. What needs some discussion, however, is the new vibrational pattern embracing the three centers.

From quantum-mechanical considerations it follows that a three-atom transition state should be linear if s electrons are involved, this conformation providing the easiest reaction path. An ordinary linear three-atom molecule will quite generally have four normal modes of vibration, the movements being illustrated in Fig. 2–2. There are two modes of linear

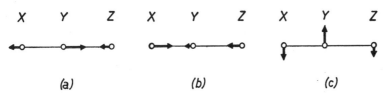

Fig. 2–2. Normal modes of vibration of a linear triatomic molecule: (a) and (b) linear, non-degenerate, (c) bending, doubly degenerate, occurring in two perpendicular planes.

vibration and two equal modes, i.e., one doubly degenerate mode, of bending vibration.

Since we need deal only with linear transition states, only two independent variables are required to describe the relative positions of the three atoms which we shall use to illustrate our three-center reaction. The two shortest internuclear distances are conveniently used as independent variables. With reference to Fig. 2–2, the symbols r_{XY} and r_{YZ} are suitable.

In order for the molecule XYZ to be a good model for our transition state, the atoms X and Z will be assigned considerably larger masses than the atom Y. The potential-energy surface, which is very useful for the present study, will in the ordinary way have the two axes r_{XY} and r_{YZ}, and the angle between them should be so chosen that the movements will be classically represented by the movements of a particle sliding on the surface under the influence of gravity. If the mass of Y is much smaller than those of X and Z, the angle will be very small (39a). For the sake of convenience, the angle has been drawn

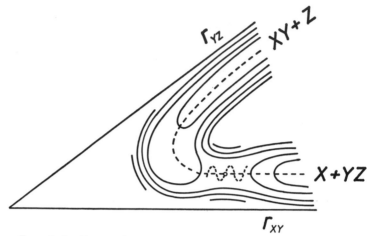

Fig. 2–3. Type of potential-energy surface of a three-atom system.

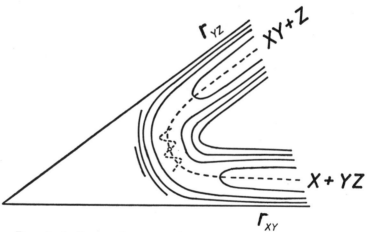

Fig. 2–4. Type of potential-energy surface of a symmetric three-atom system.

fairly large in Figs. 2–3 and 2–4 in spite of the fact that we will mostly think of the mass of Y as negligible in comparison with those of X and Z in the following, which situation corresponds to zero angle at the limit.

The following discussion attempts to facilitate the understanding of the relation between the reaction coordinate and the internuclear distances r_{XY} and r_{YZ}, which frequently causes considerable difficulty. It is necessary to make these things as clear as possible, because the (aperiodic) movement along the reaction coordinate should be excluded from the partition function of the transition state and cannot be the seat of any zero-point energy. On the other hand, there is always another vibrational mode, also corresponding to linear movements in the transition state, which will give rise to a true vibration the zero-point energy of which has to be accounted for in the partition function and which might play an important role.

We shall treat some limiting cases in which the vibrational behavior can be described in very simple terms. Our treatment will be a qualitative and descriptive one, because the corresponding exact treatment would be too lengthy and would give little more information.

In Fig. 2–3 the transition state corresponds to the same YZ distance as in the undisturbed molecule YZ, and the reaction path (the smooth broken line) is still parallel to the axis r_{XY}. At the transition state there will be a smooth valley having this direction. If we imagine a small particle moving on the surface, it is evident that its movements could be described as a superposition of an aperiodic movement along the reaction path (i.e., the reaction coordinate) and a periodic one (a true vibration) at right angles to the same direction.

Looking at the three-atom model, the first movement corresponds to a relative movement between the atom X and the molecule YZ, the two latter atoms moving at the constant normal distance from one another. This is obviously the reaction coordinate. If we start with X + YZ, the movement along the reaction path at the transition state will correspond to an approach between X and YZ; if we start with XY + Z, the movement will be in the opposite direction. The first case is pictured in Fig. 2–5; it corresponds to the vibrational mode (b) shown in Fig. 2–2.

Fig. 2–5. Aperiodic movement along the reaction coordinate in Fig. 2–3.

The transverse vibration involves a vibration of Y relative to X and Z, which keep their internuclear distance constant. Owing to their large mass, X and Z will be almost at rest. The movement is pictured in Fig. 2–6 and obviously corresponds to mode (a) in Fig. 2–2. The vibrational frequency in the limiting case is determined only by the mass of Y and of the force constant attaching it to Z; that is, the vibration behaves as in a free YZ molecule except for the magnitude of the force constant, which depends on the curvature of the valley in the direction parallel to the r_{YZ} axis.

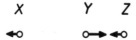

Fig. 2–6. Periodic motion, perpendicular to the reaction coordinate in Fig. 2–3.

From the chemical point of view the following conclusions could be drawn. The transition state consists of an atom X and a molecule YZ. The distance r_{XY} is the reaction coordinate, and there is a vibrational mode within YZ as if it were free, although the frequency need not be the same. Here we could say that the bond which is going to be broken in order to form XY + Z from X + YZ still has zero-point energy, and the bond which is going to be created corresponds to the reaction coordinate. If we start with XY + Z, we should say instead that the reaction coordinate in the transition state corresponds to the bond to be broken, and we already have zero-point energy in the bond to be created.

The two extremes discussed above are realized in very few reactions, if any. In most reactions the transition state will be more intermediate and not directly comparable to a free atom plus a molecule. If, instead, we approximate the situation by assuming complete symmetry, i.e., that X and Z are atoms of the same kind and have identical masses, still heavy compared to the atom Y, a very simple model is obtained if we assume that there is no basin in the potential-energy surface. The transition state will then be situated on the line bisecting the angle between the axes, and the reaction path will cut this line at right angles, as in Fig. 2–4. The reaction could be realized as some kind of hydrogen exchange between two identical heavy residues.

If a small particle is moving on the surface in Fig. 2–4, we could as before regard its movements in the vicinity of the transition state as a superposition of two normal modes, an aperiodic movement in the direction of the reaction coordinate and a periodic movement at right angles to this direction.

The movement along the reaction coordinate in the direction from X + YZ corresponds in the real system of the atomic nuclei to an aperiodic decrease in r_{XY} accompanied by an equal increase in r_{YZ}; in other words, Y goes from Z to X, the two latter atoms keeping their distance constant. In order to conserve the center of gravity of the system, Z and X are assumed to move slightly in the direction opposite that of the movement of Y, as in Fig. 2–7, which illustrates the situa-

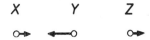

Fig. 2–7. Aperiodic movement along the reaction coordinate in Fig. 2–4.

tion. This motion obviously corresponds to the linear mode (a) of vibration of Fig. 2–2.

The periodic motion at right angles to the reaction path corresponds to movement along the line bisecting the angle between the axes. It is very easy to see that this mode of vibration is completely symmetric in X and Z, both moving outwards and inwards, the atom Y resting in the center of gravity of the system. The situation is pictured in Fig. 2–8 and corresponds to the linear mode (b) of Fig. 2–2.

Fig. 2–8. Periodic motion, perpendicular to the reaction co-ordinate in Fig. 2–4.

In this symmetric reaction the movement along the reaction coordinate at the transition state is the movement of Y relative to X and Z, the latter being fixed between themselves. The periodic symmetric motion involves zero-point energy the magnitude of which depends on the curvature of the energy surface along the bisecting line. It is obvious that both bonds could contain some zero-point energy in the transition state.

In order to estimate the influence of the isotopic mass of Y on the isotope effect, it is important to study the relation between this mass and the frequencies of the periodic motion. We shall still consider the mass of Y small relative to those of X and Z. In the unsymmetric transition state the periodic movement (Fig. 2–6) was analogous to that in a molecule YZ; thus in our approximation the frequency will be proportional to the inverse square root of the mass of atom Y, the usual dependence. In the symmetric transition state the frequency of the vibration (Fig. 2–8) is independent of the mass of the atom Y, which is resting. The zero-point energy will in the latter case be cancelled when two reactions, isotopic in Y, are compared. The doubly degenerate bending vibrations of the transition state could in both cases be approximated by motions of Y normal to the line connecting the heavy X and Z. The frequencies are proportional to the inverse square root of the mass of Y in the usual manner. To sum up, the zero-point energy that shows any dependence on the mass of Y, and thus is not cancelled in the isotope effect, will be proportional to its inverse square root. Thus Eq. (2–13) is still applicable.

It is evident that all approximations are very rough, because the formation of a single molecule from two not too simple molecules, X and YZ, involves the formation of six new internal degrees of freedom, and, if an atom X is added to such a molecule YZ, three new degrees arise. The same holds, of course, if YZ is an activated complex. The above discussion in terms of a three-atom model is unable to account for all of these and for the influence of isotopic mass on the enlarged vibrational pattern.

Let us, however, try to compare the three-center reaction with the two-center one, in which the stretching mode of the bond to be broken constitutes the reaction coordinate. If the isotopic difference between the two isotopic reactions is limited to the single transferred atom Y of very small mass, the contributions from the reactants to the exponent of Eq. (2–8) of a model consisting of the dissociation of YZ alone do not differ from those of a model containing an acceptor X, because X is not isotopic and all $u_{(1)} = u_{(2)}$. As usual, the contributions come from the three vibrational modes of the Y—Z bond (Z has considerable moments of inertia). In the transition state the two-center model has lost its stretching vibration but has still two modes of bending, which might be accompanied by large or small zero-point energies. The three-center model has two similar modes of bending vibration and in addition a new mode of linear vibration. The difference is, in principle, limited to this extra linear vibration. Probably most reactions are closer to the symmetric case than to the unsymmetric one, both having been discussed above. If the reaction is also fairly symmetric in the masses of X and Z, the contribution of this vibration to the exponent in Eq. (2–8) will be negligible because $u_{(1)}{}^{\ddagger} \approx u_{(2)}{}^{\ddagger}$. Thus Eqs. (2–10) and (2–12) are probably still applicable as approximations. It should also be observed that, if the unsymmetric model of the transition state is a better approximation, the new linear vibrational mode tends to weaken the isotope effect via the zero-point energy of the former if the frequency is high, because

$$u_{(\text{heavy})}{}^{\ddagger} < u_{(\text{light})}{}^{\ddagger}$$

If the frequency is very low, the zero-point energy is negligible but the vibrational partition function introduces the factor $u_{(2)}{}^{\ddagger}/u_{(1)}{}^{\ddagger}$, which also tends to weaken the isotope effect.

Introduction of an acceptor thus might weaken the isotope effect as explained above, but the influence on the bending vibrations was then neglected. It seems very probable that the bending frequencies might be considerably increased. As usual, increase in the transition state of zero-point energies sensitive to the isotopic mass tends to weaken the isotope effect. Therefore the presence of an acceptor seems, in general, to weaken the isotope effect. For approximate calculations, however, we can do little but use Eqs. (2–10) and (2–12).

Secondary Hydrogen Isotope Effects. If the isotopic hydrogen is located at some position of the molecule other than the direct reaction center and is not split off in the reaction, the frequencies indexed (1) and (2) in Eqs. (2–10), (2–11), and (2–12) do not belong to atoms of different mass; the index is merely a reminder of the existence of different isotopes in some other position. m_1 and m_2 in Eqs. (2–11) and (2–12) will, of course, also be equal. To a first approximation, the frequencies $\nu_{k(1)}$ and $\nu_{k(2)}$ are equal, and no isotope effect arises. The frequencies of those modes of vibration, which are sensitive to the isotopic differences, are, according to our present picture, assumed to be cancelled between the two sums of the last exponent of Eq. (2–8). This is in agreement with the experimental experience that secondary isotope effects will generally be small and frequently negligible. In the cases known, where secondary isotope effects play an important role, this is generally ascribed to the fact that passing into the transition state influences the binding forces of neighboring or remote bonds to such an extent that there is no longer any pair-wise matching between those frequencies in the transition state and in the reactant which are sensitive to the isotopic differences (cf. Chapter 5).

BIGELEISEN'S TREATMENT

With few exceptions, an expression of the type of Eq. (2–5) is not very convenient in the case of isotopes heavier than hydrogen. The zero-point energy will then no longer play such a dominant role as with hydrogen. For instance, the deviations in the moments of inertia will be of the same order of magnitude as the deviations of the final rate-constant ratio from unity. It is no longer possible to neglect the influence of any factor in Eq. (2–5).

Bigeleisen's Complete Expression. The difficulty can be overcome by replacing the ratios of molecular masses and moments of inertia by the ratios of the individual atomic masses (which will disappear by cancellation) and the ratios of the frequencies of the different modes of vibration (which might be partly known and partly estimated). Thus the isotopic influence on the molecular magnitudes is concentrated into the influence on the vibrational frequencies. According to the product theorem of Teller and Redlich (75), the following relation holds for two ordinary isotopic molecules:

$$\left(\frac{M_1}{M_2}\right)^{3/2} \left(\frac{A_1 B_1 C_1}{A_2 B_2 C_2}\right)^{1/2} = \prod_{j}^{n}\left(\frac{m_{j(1)}}{m_{j(2)}}\right)^{3/2} \prod_{i}^{3n-6} \frac{\nu_{i(1)}}{\nu_{i(2)}} \qquad (2\text{--}14)$$

where m denotes atomic mass and n is the number of atoms. The products are to be taken over all atoms or frequencies, respectively. For linear molecules the number of vibrations should be increased by one, and the moments of inertia decreased from three to two (which are equal).

In the transition state one mode of vibrational motion has an imaginary frequency corresponding to the motion along the decomposition coordinate. This frequency is called $\nu_L^{\ddagger}$. Consequently, for the transition state the following holds:

$$\left(\frac{M_1^{\ddagger}}{M_2^{\ddagger}}\right)^{3/2} \left(\frac{A_1^{\ddagger} B_1^{\ddagger} C_1^{\ddagger}}{A_2^{\ddagger} B_2^{\ddagger} C_2^{\ddagger}}\right)^{1/2} = \frac{\nu_{L(1)}^{\ddagger}}{\nu_{L(2)}^{\ddagger}} \prod_{j}^{n\ddagger}\left(\frac{m_{j(1)}^{\ddagger}}{m_{j(2)}^{\ddagger}}\right)^{3/2} \prod_{i}^{3n\ddagger-7} \frac{\nu_{i(1)}^{\ddagger}}{\nu_{i(2)}^{\ddagger}} \qquad (2\text{--}15)$$

Equations (2–14) and (2–15) can be substituted directly in the primary expression (2–5). It should be observed that $\nu_L^{\ddagger}$ corresponds to the vibrational mode which was excluded from the partition functions. With $u = h\nu/kT$ as usual, we obtain

$$\frac{k_1}{k_2} \times \frac{s_2}{s_1} \times \frac{s_1^{\ddagger}}{s_2^{\ddagger}} =$$

$$= \frac{\nu_{L(1)}^{\ddagger}}{\nu_{L(2)}^{\ddagger}} \prod_{j}^{n\ddagger}\left(\frac{m_{j(1)}^{\ddagger}}{m_{j(2)}^{\ddagger}}\right)^{3\!/\!2} \prod_{i}^{3n\ddagger-7}\left(\frac{u_{i(1)}^{\ddagger}}{u_{i(2)}^{\ddagger}} \times \frac{e^{-\frac{1}{2}u_{i(1)}^{\ddagger}}}{e^{-\frac{1}{2}u_{i(2)}^{\ddagger}}} \times \frac{1-e^{-u_{i(2)}^{\ddagger}}}{1-e^{-u_{i(1)}^{\ddagger}}}\right) \times$$

$$\times \prod_{j}^{n}\left(\frac{m_{j(2)}}{m_{j(1)}}\right)^{3\!/\!2} \prod_{i}^{3n-6}\left(\frac{u_{i(2)}}{u_{i(1)}} \times \frac{e^{-\frac{1}{2}u_{i(2)}}}{e^{-\frac{1}{2}u_{i(1)}}} \times \frac{1-e^{-u_{i(1)}}}{1-e^{-u_{i(2)}}}\right)$$

It is easy to see that transition state (1) and reactant (2) together contain exactly the same set of atoms as transition state (2) and reactant (1) together. Hence all atomic masses will be cancelled, and we are left with

$$\frac{k_1}{k_2} \times \frac{s_2}{s_1} \times \frac{s_1^{\ddagger}}{s_2^{\ddagger}} = \frac{\nu_{L(1)}^{\ddagger}}{\nu_{L(2)}^{\ddagger}} \prod_{i}^{3n\ddagger-7}\left(\frac{u_{i(1)}^{\ddagger}}{u_{i(2)}^{\ddagger}} \times \frac{e^{-\frac{1}{2}u_{i(1)}^{\ddagger}}}{e^{-\frac{1}{2}u_{i(2)}^{\ddagger}}} \times \frac{1-e^{-u_{i(2)}^{\ddagger}}}{1-e^{-u_{i(1)}^{\ddagger}}}\right) \times$$

$$\times \prod_{i}^{3n-6}\left(\frac{u_{i(2)}}{u_{i(1)}} \times \frac{e^{-\frac{1}{2}u_{i(2)}}}{e^{-\frac{1}{2}u_{i(1)}}} \times \frac{1-e^{-u_{i(1)}}}{1-e^{-u_{i(2)}}}\right) \qquad (2\text{–}16)$$

This expression is equivalent to Eq. (2–5), and it could be called Bigeleisen's complete expression (10, 14).* The ratio $\nu_{L(1)}^{\ddagger}/\nu_{L(2)}^{\ddagger}$ will be discussed below. The large products could be said to represent the ratio of the complete activation equilibrium constants, except for the symmetry numbers. A comparison of Eqs. (2–16) and (2–5), neglecting symmetry numbers, shows that the product P of the two large products can be written

$$P = \frac{\nu_{L(2)}^{\ddagger}}{\nu_{L(1)}^{\ddagger}} \times \frac{K_1^{\ddagger}}{K_2^{\ddagger}} = \frac{\dfrac{kT}{h\nu_{L(1)}^{\ddagger}} K_1^{\cdot}}{\dfrac{kT}{h\nu_{L(2)}^{\ddagger}} K_2^{\ddagger}}$$

Comparison with Eq. (2–2) justifies the above statement. P could also be said to represent the equilibrium constant of the isotope exchange equilibrium

* Equation (2–16) could be called a "primary" expression, just as Eq. (2–5) is. This is evident if the former is derived by means of corrections for quantum-mechanical vibrational behavior applied to the classical system (12) as originally done by Bigeleisen (10).

(reactant)$_1$ + (transition state)$_2$ $\rightleftharpoons$

(reactant)$_2$ + (transition state)$_1$

That this should be so is also evident from the work of Bigeleisen and Goeppert Mayer (12). Any vibration that behaves classically causes no isotope fractionation, and hence it is immaterial whether it is included or not in the products of Eq. (2–16). If in setting up the equilibrium constant directly the missing u of the transition state is included, nothing is changed, because a vanishing u contributes nothing but a factor of unity.

Some General Approximations. The relation between Eq. (2–16) and the approximate equations derived above is easily seen as follows. If all real frequencies except the one being approximately assigned to the cleaving bond (ν_k) are assumed to be cancelled as discussed on p. 16, Eq. (2–16) simplifies to

$$\frac{k_1}{k_2} \times \frac{s_2}{s_1} \times \frac{s_1^{\ddagger}}{s_2^{\ddagger}} = \frac{\nu_{L(1)}^{\ddagger}}{\nu_{L(2)}^{\ddagger}} \times \frac{u_{k(2)}}{u_{k(1)}} \times \frac{e^{-\frac{1}{2}u_{k(2)}}}{e^{-\frac{1}{2}u_{k(1)}}} \times \frac{1 - e^{-u_{k(1)}}}{1 - e^{-u_{k(2)}}}$$

Since $\nu_L^{\ddagger}$ and ν_k represent more or less corresponding motions of the transition state and the reactant, respectively, it seems natural to assign the value unity to the product of the two first ratios to the right, obtaining

$$\frac{k_1}{k_2} \times \frac{s_2}{s_1} \times \frac{s_1^{\ddagger}}{s_2^{\ddagger}} = \frac{\sinh \frac{1}{2}u_{k(1)}}{\sinh \frac{1}{2}u_{k(2)}}$$

which is Eq. (2–7), proposed by Eyring and Cagle (35). High u_k values make this equivalent to Eq. (2–10) also.

The case with isotopic hydrogen discussed on pp. 21–22 assumes cancellation of all real frequencies except the three approximately ascribed to the heavy-atom-hydrogen bond in the reactant and the two corresponding bending modes of the transition state. The three former frequencies are assumed to be large, the two latter vanishing. This gives

$$\frac{k_1}{k_2} \times \frac{s_2}{s_1} \times \frac{s_1^{\ddagger}}{s_2^{\ddagger}} = \frac{\nu_{L(1)}^{\ddagger}}{\nu_{L(2)}^{\ddagger}} \prod_i^3 \frac{u_{i(2)}}{u_{i(1)}} \times \exp\left\{ \frac{1}{2} \sum_i^3 (u_{i(1)} - u_{i(2)}) \right\}$$

since

$$\lim_{u_i\ddagger \to 0} \frac{1 - e^{-u_{i(2)}\ddagger}}{1 - e^{-u_{i(1)}\ddagger}} = \frac{u_{i(2)}\ddagger}{u_{i(1)}\ddagger}$$

The imaginary frequency and the real stretching one are cancelled as above, and the two bending frequency ratios together give m_1/m_2; hence

$$\frac{k_1}{k_2} \times \frac{s_2}{s_1} \times \frac{s_1\ddagger}{s_2\ddagger} = \frac{m_1}{m_2} \exp \left\{ \frac{1}{2} \sum_i^3 (u_{i(1)} - u_{i(2)}) \right\}$$

which is Eq. (2–12).

From the physical point of view it is interesting, but hardly astonishing, to observe that the assumption of a cancellation of the molecular masses and the moments of inertia is about equivalent to the assumption that most frequency ratios are cancelled. The possibility of assigning definite frequencies to the hydrogen-heavy-residue oscillator is confirmed by the general experience from infrared spectroscopy. This means also that the vibrational behavior of the "heavy residue" is mainly independent of the isotopic mass of the hydrogen. It is obvious that our "residue" must be "heavy" if it is possible to assign the value $(m_1/m_2)^{\frac{1}{2}}$ to the stretching $u_{(2)}/u_{(1)}$. In the same way it must be rotationally "inert" if the same value holds also for each bending $u_{(2)}/u_{(1)}$. Hence neither the molecular masses nor the moments of inertia could be appreciably sensitive to the isotopic mass of the hydrogen.

Heavy-Atom Approximation. The following approximations, which are useful with carbon and heavier elements, were introduced initially by Bigeleisen and Goeppert Mayer (12) for the case of isotope exchange equilibria and have been carried over to reaction rates by Bigeleisen (10).

If the substitution $u_{(1)} - u_{(2)} = \Delta u$ is introduced in Eq. (2–16) and the natural logarithm of each side is taken, we have

$$\ln \frac{k_1}{k_2} \times \frac{s_2}{s_1} \times \frac{s_1\ddagger}{s_2\ddagger} =$$

$$= \ln \frac{\nu_{L(1)}\ddagger}{\nu_{L(2)}\ddagger} + \sum_i^{3n\ddagger-7} \left[\ln \left(1 + \frac{\Delta u_i\ddagger}{u_{i(2)}\ddagger} \right) - \frac{1}{2} \Delta u_i\ddagger + \ln \frac{1 - e^{-u_{i(2)}\ddagger}}{1 - e^{-(u_{i(2)}\ddagger + \Delta u_i\ddagger)}} \right] -$$

$$- \sum_{i}^{3n-6} \left[\ln \left(1 + \frac{\Delta u_i}{u_{i(2)}} \right) - \frac{1}{2} \Delta u_i + \ln \frac{1 - e^{-u_{i(2)}}}{1 - e^{-(u_{i(2)}+\Delta u_i)}} \right] \qquad (2\text{-}17)$$

For carbon and heavier elements, Δu is small compared to u and, at not too low temperatures, also compared to unity. This allows the following approximations:

$$\ln \left(1 + \frac{\Delta u}{u} \right) - \frac{1}{2} \Delta u + \ln \frac{1 - e^{-u}}{1 - e^{-(u+\Delta u)}} \approx$$

$$\approx \frac{\Delta u}{u} - \frac{\Delta u}{2} + \ln \frac{e^u - 1}{e^u - e^{-\Delta u}} =$$

$$= \frac{\Delta u}{u} - \frac{\Delta u}{2} - \ln \frac{e^u - 1 + \Delta u - \cdots}{e^u - 1} \approx$$

$$\approx \frac{\Delta u}{u} - \frac{\Delta u}{2} - \ln \left(1 + \frac{\Delta u}{e^u - 1} \right) \approx$$

$$\approx \left(\frac{1}{u} - \frac{1}{2} - \frac{1}{e^u - 1} \right) \Delta u \equiv -G(u) \, \Delta u$$

With the function $G(u)$ defined as above, Eq. (2-17) may be written approximately

$$\ln \frac{k_1}{k_2} \times \frac{s_2}{s_1} \times \frac{s_1^{\ddagger}}{s_2^{\ddagger}} = \ln \frac{\nu_{L(1)}^{\ddagger}}{\nu_{L(2)}^{\ddagger}} - \sum_{i}^{3n\ddagger-7} G(u_i{}^{\ddagger}) \, \Delta u_i{}^{\ddagger} + \sum_{i}^{3n-6} G(u_i) \, \Delta u_i$$

$$(2\text{-}18)$$

Since the two sums are small compared to unity, it is possible to take the antilogarithms in the following way:

$$\frac{k_1}{k_2} \times \frac{s_2}{s_1} \times \frac{s_1^{\ddagger}}{s_2^{\ddagger}} = \frac{\nu_{L(1)}^{\ddagger}}{\nu_{L(2)}^{\ddagger}} \left(1 - \sum_{i}^{3n\ddagger-7} G(u_i{}^{\ddagger}) \, \Delta u_i{}^{\ddagger} + \sum_{i}^{3n-6} G(u_i) \, \Delta u_i \right)$$

$$(2\text{-}19)$$

Equation (2-19) is the form of the expression commonly used for the computation of carbon and other heavy-atom isotope effects. The approximations made in its derivation are not valid in the case of hydrogen.

The function $G(u)$ has been conveniently tabulated by Bigeleisen and Goeppert Mayer (12), and the table is also found in the Appendix of Dole's *Introduction to Statistical Thermodynamics* (32).

For convenient approximate methods of including the tunneling correction from Eq. (2–1) into this formalism, and of using force-constant data in a simplified manner, see the recent review by Bigeleisen and Wolfsberg (14).

The Factor $\nu_{L(1)}{}^{\ddagger}/\nu_{L(2)}{}^{\ddagger}$. Hitherto fairly little has been said about the temperature-independent factor $\nu_{L(1)}{}^{\ddagger}/\nu_{L(2)}{}^{\ddagger}$. In the earlier literature it was generally given as the equivalent $(m_2^*/m_1^*)^{1/2}$, where m^* denotes the reduced mass along the decomposition coordinate. Unfortunately the computation of this factor introduces considerable uncertainty, which, however, does not influence the temperature dependence. The problem has been reviewed recently by Bigeleisen and Wolfsberg (14).

In earlier work, m^* was generally replaced by the reduced mass of the two atoms joined by the bond being broken. This being in agreement with work according to classical models by Slater, a decomposition coordinate treated in this manner is frequently referred to as a Slater coordinate. If the two atomic masses are m' and m'', we have

$$\frac{\nu_{L(1)}{}^{\ddagger}}{\nu_{L(2)}{}^{\ddagger}} = \left(\frac{m_2^*}{m_1^*}\right)^{1/2} = \left[\frac{\dfrac{1}{m_1'} + \dfrac{1}{m_1''}}{\dfrac{1}{m_2'} + \dfrac{1}{m_2''}}\right]^{1/2} \tag{2–20}$$

More recently the reduced mass has been calculated in the same manner but with the masses of the separating fragments instead of those of the separating atoms. If the molecular fragments have the masses M' and M'', the temperature-independent factor becomes

$$\frac{\nu_{L(1)}{}^{\ddagger}}{\nu_{L(2)}{}^{\ddagger}} = \left(\frac{m_2^*}{m_1^*}\right)^{1/2} = \left[\frac{\dfrac{1}{M_1'} + \dfrac{1}{M_1''}}{\dfrac{1}{M_2'} + \dfrac{1}{M_2''}}\right]^{1/2} \tag{2–21}$$

Bigeleisen and Wolfsberg (13) have given a particular formula for the temperature-independent factor to be used in the case of three-center reactions. The reaction coordinate x_L corresponding to the reacting system

$$X + YZ \longrightarrow XY + Z$$

depends on the distances r_{XY} and r_{YZ} between the reacting atoms (or fragments) and can be written

$$x_L = \alpha r_{YZ} - \beta r_{XY}$$

where α and β are parameters. For a symmetric transition state like the one in Fig. 2–4, α and β are equal, and the forward motion along x_L involves an increase in r_{YZ} and an equal decrease in r_{XY}. If the transition state is of the type pictured in Fig. 2–3, on the other hand, the motion along the reaction coordinate involves simply a decrease in r_{XY}; hence $\alpha = 0$, $\beta = 1$. If, finally, we have the mirror situation of Fig. 2–3, i.e., if the reaction coordinate is merely the increasing distance r_{YZ}, $\alpha = 1$, $\beta = 0$.

Defining $p = \beta^2/\alpha^2$, the values of p in the three examples mentioned above are 1, ∞, and 0, respectively. Using the parameter p, Bigeleisen and Wolfsberg give the expression

$$\frac{\nu_{L(1)}^{\ddagger}}{\nu_{L(2)}^{\ddagger}} = \left[\frac{\dfrac{1}{m_{Y(1)}} + \dfrac{1}{m_{Z(1)}} + p\left(\dfrac{1}{m_{Y(1)}} + \dfrac{1}{m_{X(1)}}\right) + 2p^{\frac{1}{2}}\dfrac{1}{m_{Y(1)}}}{\dfrac{1}{m_{Y(2)}} + \dfrac{1}{m_{Z(2)}} + p\left(\dfrac{1}{m_{Y(2)}} + \dfrac{1}{m_{X(2)}}\right) + 2p^{\frac{1}{2}}\dfrac{1}{m_{Y(2)}}} \right]^{\frac{1}{2}}$$

$$(2\text{--}22)$$

for a linear transition state.

It is evident that the situation defined by $p = 0$ corresponds to the same temperature-independent ratio as a pure decomposition of YZ, and that defined by $p = \infty$ to the same ratio as the simple combination of X and Y (which ratio is, of course, the same as for the corresponding simple decomposition).

The Bigeleisen Formalism in Practice. It should be observed that $G(0) = 0$; that is, a vanishing frequency gives no contribution to the sums of Eq. (2–19). That this should be so was pointed out in the discussion of Eq. (2–16). From the discussions of three-center reactions it should be evident that the relation between the reaction coordinate and the interatomic distances is very complex in the general case. On the other hand, one is frequently forced to discuss the vibrational behavior of a molecule in terms of "bond vibrations." By definition, the movement along the reaction coordinate is not associated with zero-point energy, but the other normal mode of linear vibration of a transition state certainly is. This is frequently difficult to translate into "bond vibrations." Expressions like "the bond is assigned zero frequency in the transition state," frequently met in the literature, might mean either that the frequency is low but real or that the corresponding motion is identified with the motion along the decomposition coordinate. The effect on the temperature-dependent factor is the same. From Eqs. (2–16) and (2–19), however, it is evident that the product or sum, respectively, of the transition state should contain one vibration less than in the case of an ordinary molecule. Since the vibrational pattern is still not resolved for most stable molecules, and so much less for transition states, it will generally be necessary to match most of the vibrational frequencies pair-wise between each reactant and its transition state or between two isotopic species of the same kind, as was done in practice with Eq. (2–5). In the simplest cases one vibration of the reactant is finally left, as shown on p. 35.

Occasionally it might be better, even for heavy elements, to calculate the approximate moments of inertia (particularly for rigid entities) and use Eq. (2–6) instead of Eq. (2–16) or (2–19), with the assumption that all frequencies but one (u_k) of the reactant are cancelled. The reason is that the moments of inertia might introduce less uncertainty than the oversimplified frequency ratios together with the ratio of the reduced masses along the decomposition coordinate (cf. pp. 147 ff.).

The particular value of Bigeleisen's treatment depends on its use on one hand with small molecules the vibrational pattern of which is known in detail, and on the other with all non-rigid molecules containing isotopes so heavy that the zero-point energy is no longer decisive.

INTERRELATION OF DIFFERENT THEORETICAL EXPRESSIONS

The plan on p. 42 is a survey of the different expressions for the computation of isotopic rate-constant ratios and of their interrelations. Each box corresponds to an expression, generally given in this text, and at the arrows the assumptions underlying the transformation of one formula into another are indicated. Descending in this scheme means making use of simplifying assumptions. The equations referred to below the level of the Eyring-Cagle formula are, in general, applicable only to isotopic hydrogen. M denotes molecular masses in general, ABC moments of inertia in general, and u, as usual, $h\nu/kT$.

ACTIVITY COEFFICIENTS

Hitherto little has been said about the activity coefficients, which occur in the complete absolute reaction rate expression (2–1). The ratio $\alpha_A \alpha_B \cdots / \alpha^{\ddagger} \alpha_N \cdots$ of properly defined activity coefficients serves to transfer the activation equilibrium from the conditions for which $K^{\ddagger}$ was computed to other conditions; that is, the activity coefficient ratio accounts for the change in the standard free energy of activation due to external influence (39b). In the calculation of isotopic rate-constant ratios only the relative change in $\alpha_A \alpha_B \cdots / \alpha^{\ddagger} \alpha_N \cdots$, or rather in $\alpha_A / \alpha^{\ddagger}$, with isotopic substitution need be known. In other words, the activity coefficients introduce an extra isotope effect only if the isotopic substitution causes a change in the external contribution to the free energy of activation.

The method of computation of $K^{\ddagger}$ from statistical mechanics assumes fairly ideal conditions, which are likely to prevail only in the gas phase, where the molecules are far apart. Most experiments, on the other hand, are carried out in the liquid

Interrelation of Different Theoretical Expressions

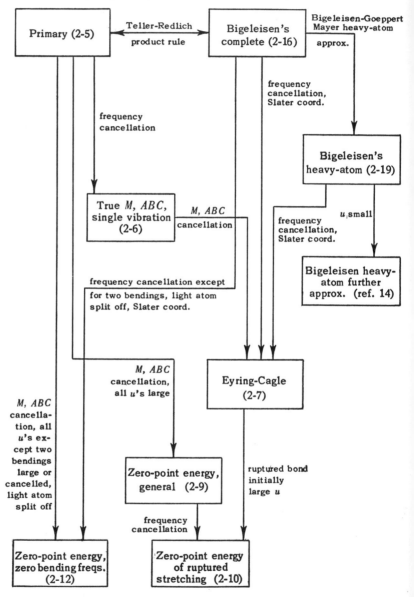

phase, where perturbing influences are strong. It is, however, generally assumed that the external influences have little isotopic specificity. The justification comes from the known fact that isotopic substitution has fairly little influence on solubilities and other properties related to interactions between a molecule and its environment. It must be admitted, however, that little attention has been given the problem hitherto. The problem is touched upon on p. 153.

The strongest effects arising from interactions with the medium could be expected with the hydrogen isotopes. A strong interaction, likely to be subject to isotope effects, is represented by hydrogen-bond formation. Unfortunately, however, the same conditions favor hydrogen-bond formation and hydrogen exchange, and the latter renders isotope effect studies impossible.

In the following, little will be said about activity coefficients, and this involves an implicit assumption that the interaction with the medium does not introduce an additional isotope effect.

TEMPERATURE DEPENDENCE OF ISOTOPE EFFECTS

Low Temperature. From Eq. (2–5) or (2–16) we easily obtain the low-temperature limit of the isotope effect. For very large u's, the following relations hold respectively:

$$\frac{k_1}{k_2} \times \frac{s_2}{s_1} \times \frac{s_1^\ddagger}{s_2^\ddagger} = \left(\frac{M_1^\ddagger}{M_2^\ddagger} \times \frac{M_2}{M_1}\right)^{3/2} \left(\frac{A_1^\ddagger B_1^\ddagger C_1^\ddagger}{A_2^\ddagger B_2^\ddagger C_2^\ddagger} \times \frac{A_2 B_2 C_2}{A_1 B_1 C_1}\right)^{1/2} \times$$

$$\times \exp\left\{-\frac{1}{2}\left[\sum_i^{3n\ddagger-7}(u_{i(1)}^\ddagger - u_{i(2)}^\ddagger) - \sum_i^{3n-6}(u_{i(1)} - u_{i(2)})\right]\right\}$$

$$(2\text{–}23)$$

and

$$\frac{k_1}{k_2} \times \frac{s_2}{s_1} \times \frac{s_1^\ddagger}{s_2^\ddagger} = \frac{\nu_{L(1)}^\ddagger}{\nu_{L(2)}^\ddagger} \prod_i^{3n\ddagger-7} \frac{u_{i(1)}^\ddagger}{u_{i(2)}^\ddagger} \prod_i^{3n-6} \frac{u_{i(2)}}{u_{i(1)}} \times$$

$$\times \exp\left\{-\frac{1}{2}\left[\sum_{i}^{3n\ddagger-7}(u_{i(1)}{}^{\ddagger}-u_{i(2)}{}^{\ddagger})-\sum_{i}^{3n-6}(u_{i(1)}-u_{i(2)})\right]\right\}$$

$$(2\text{--}24)$$

Depending on the sign of the zero-point-energy difference, the ratio of the rates will tend toward zero or infinity when T becomes infinitely small. Generally, the lighter of two isotopic molecules will have the largest zero-point energy, and the reactant will have larger zero-point energy than the corresponding isotopic form of the transition state. From this it follows that the light molecule should react infinitely faster than the heavy one at $T = 0$, provided possible complications introduced by the transmission coefficients need not be taken into account.

In the region of low temperatures the difference in experimental activation energy between isotopic reactions will acquire a particularly simple form. Denoting the activation energy (in molar amount) by E, we have, from Eq. (2–23) or (2–24),

$$E_1 - E_2 = RT^2\frac{d\ln(k_1/k_2)}{dT} =$$

$$= -\frac{RT^2}{2}\times\frac{d}{dT}\left[\sum_{i}^{3n\ddagger-7}(u_{i(1)}{}^{\ddagger}-u_{i(2)}{}^{\ddagger})-\sum_{i}^{3n-6}(u_{i(1)}-u_{i(2)})\right] =$$

$$= \frac{Nhc}{2}\left[\sum_{i}^{3n\ddagger-7}(\bar{\nu}_{i(1)}{}^{\ddagger}-\bar{\nu}_{i(2)}{}^{\ddagger})-\sum_{i}^{3n-6}(\bar{\nu}_{i(1)}-\bar{\nu}_{i(2)})\right] \qquad (2\text{--}25)$$

The low-temperature difference in activation energy is thus simply equal to the difference between the transition state and the reactant of the differences in zero-point energy between the isotopic species. That this should be so at low temperatures, when all oscillators have gathered in the lowest levels, is evident from Fig. 2–1.

High Temperature. The limit at high temperatures, i.e., for vanishing u's, is most easily obtained from Eq. (2–16). The products approach unity, leaving

$$\frac{k_1}{k_2} \times \frac{s_2}{s_1} \times \frac{s_1^{\ddagger}}{s_2^{\ddagger}} = \frac{\nu_{L(1)}^{\ddagger}}{\nu_{L(2)}^{\ddagger}} \qquad (2\text{--}26)$$

From the general principles of isotope exchange equilibria, it is known that classical behavior gives rise to no equilibrium isotope fractionation (12). Thus at high temperatures, when all systems approach classical behavior, the equilibrium between transition state and reactant causes no isotope fractionation. The decomposition rate of the transition state is the single source of an isotope effect, and the ratio is as above. As usual, the light molecule will react faster than the heavy one.

Since at high temperatures the rate-constant ratio is temperature-independent, there will be no difference in experimental activation energy between isotopic molecules.

Ordinary Temperatures. At room temperature the behavior of most ordinary chemical systems is much closer to the low-temperature approximations of Eq. (2–23) or (2–24) than to the high-temperature approximation of Eq. (2–26), because the picture with all oscillators in their zero-point levels is a much better model than the picture with a classical partition function for the oscillators.

If desired, the difference in experimental activation energy at intermediate temperatures could be calculated from any of the appropriate general expressions of the rate-constant ratio by derivation, according to the well-known formula

$$E_1 - E_2 = RT^2 \frac{d \ln (k_1/k_2)}{dT}$$

The pertinent relations for Bigeleisen's treatment are found in his well-known paper (10), published in 1949.

3

Evaluation of Rate-Constant Ratios from Experimental Data

If macroscopic amounts of preparations of isotopically completely pure molecules are available, and the reaction is carried through with each isotopic form separately, it is in principle a simple task to measure the corresponding specific rates and hence the isotope effect, once the kinetics of the reaction have been disentangled. In many cases, however, it is impossible or too difficult to prepare isotopically completely pure substances, and, moreover, most isotopes are not available in complete purity. In such cases it would frequently not be economical to make an effort to obtain a very high isotopic purity, because valuable results could often be obtained with one isotope present only in tracer amounts. Moreover, with radioactive isotopes it is hardly desirable to work with too high specific activities. Generally it is better to choose the specific activity which will be convenient for the method of isotope analysis concerned, usually a fairly low specific activity, which does not require too much precaution against radiation.

With a mixture of isotopic molecules, such as will always be present in experiments with tracer quantities of one of the isotopes, a difference in reaction rate between the components will always cause the composition of the reactant as well as that of the product to change during the course of the reaction. It will generally be necessary to ascertain the extent of the reaction before the rate-constant ratio can be evaluated from measurements of the isotopic composition of the initial material and that of the product or the remaining reactant. This evaluation usually requires some care but involves, on the other hand, no particular difficulty once all different parallel reactions of the isotopic species have been duly accounted for.

In this chapter little will be said about the reaction mechanism. The reactions will generally be supposed to be unimolecular in the isotopic reactant, which is the case with the majority of more complex reactants. Whether the reaction is a single-step one or the result of a complex stepwise process (or even if it is really rate-determining for the over-all reaction) is immaterial, once it is rate-determining from the point of view of the isotopic reactant.

The relation between the mechanism of complex reactions and the strength of the isotope effect will be treated in Chapters 6 and 8. It should be stressed here, however, that the isotope effect observed in separate isotopic runs need not be the same as that observed in competitive experiments if the reaction is complex.

The problems discussed in this chapter have also been analyzed by Bigeleisen and Wolfsberg in their recent review (14).

INTRAMOLECULAR COMPETITION ONLY

The kinetic situation which is simplest to treat is the one with competition only within the same molecule. If the reactant of interest from the "isotopic" point of view consists of a single kind of molecule with respect to the isotopic composition also but has several chemically equivalent positions, which will compete, the reaction scheme for a reaction unimolecular in PA_{m+n} is simply

$$PA_{(1)m}A_{(2)n} \left\{ \begin{array}{l} \xrightarrow{k'} \ RA_{(1)} + \cdots \\ \xrightarrow{k''} \ RA_{(2)} + \cdots \end{array} \right.$$

The two isotopes will always react in the ratio k'/k'' irrespectively of how the concentration of the isotopic reactant changes, and $RA_{(1)}$ and $RA_{(2)}$ will be formed in this constant ratio throughout the reaction. All calculations of the ratio k'/k'' from experimental data are self-evident.

The extreme simplicity of the isotope kinetics makes this kind of study convenient from the computational point of view. On the other hand, the synthesis of samples perfectly pure isotopically is sometimes very difficult. If the sample consists of only two kinds of isotopic molecules, however, it might be possible to follow the conversion of each of the isotopic species separately provided that the isotopic compositions of the initial and recovered reactant, as well as that of the product, are available for determination. Such experiments have been carried out by Wiberg and Slaugh (102); see also p. 73 in the present volume.

It should be observed that the results of measurements of the present kind always represent intramolecular isotope effects. When a light and a heavy isotope compete, the rate of the light one might be subject to secondary isotope effects caused by the presence of the heavy atom in another part of the molecule. The reaction rate of the light isotope might consequently differ somewhat from that in an all-light molecule.

COMPETITION BETWEEN TWO ISOTOPIC MOLECULAR SPECIES

If two isotopic species $A_{(1)}$ and $A_{(2)}$ react with different specific rates with the non-isotopic species B, C, etc., we have the following scheme:

$$A_{(1)} + B + C + \cdots \xrightarrow{k_1} (\text{products})_1$$

$$A_{(2)} + B + C + \cdots \xrightarrow{k_2} (\text{products})_2$$

The scheme is applicable to all reactions which are unimolecular in A. Even if this were not the case, but if one of the isotopic species, say $A_{(1)}$, were present only in tracer amounts, the scheme would still be applicable if, for instance, B were put equal to $A_{(2)}$ in both lines. This is to say that a reaction will practically never be bimolecular in a tracer molecule. Such coincidences will be entirely negligible. The second molecule of A will then be the same in all reactions of practical importance and can thus be considered non-isotopic in this treatment. The amount of $A_{(2)}$ consumed via transition states in which $A_{(1)}$ is also involved can be neglected and will therefore not interfere with the determination of k_2.

If the general rate expression of the (irreversible) reaction could be factorized in the following way,

$$v = -\frac{da}{dt} = k \times a \times f (b, c, \cdots)$$

(a, b, c, etc., denote concentration of the respective species), as can frequently be done with ordinary homogeneous reactions, we obtain simply

$$\frac{1}{k_1} \times \frac{da_1}{a_1} = \frac{1}{k_2} \times \frac{da_2}{a_2}$$

This expression might be integrated at once. With the initial conditions $a_1 = a_1^0$ and $a_2 = a_2^0$ we obtain

$$\frac{k_1}{k_2} = \frac{\log (a_1/a_1^0)}{\log (a_2/a_2^0)} \tag{3-1}$$

The amount of reaction of isotopic species 1 is denoted by x_1 and that of species 2 by x_2. Since, for instance, $a_1/a_1^0 = 1 - x_1$, we obtain

$$\frac{k_1}{k_2} = \frac{\log (1 - x_1)}{\log (1 - x_2)} \tag{3-2}$$

or

$$x_1 = 1 - (1 - x_2)^{k_1/k_2} \tag{3-3}$$

This function is shown in Fig. 3–1.

From a study of the function (3–3) it is evident that the fractional conversions of the isotopic species will initially be in the ratio of their specific rates. At the end of the conversion there will generally be a considerable lagging behind by the slow-reacting isotopic species if the specific rates are fairly different. This important feature of the function close to the point $x_1 = 1$, $x_2 = 1$ will be evident if we rewrite Eq. (3–3):

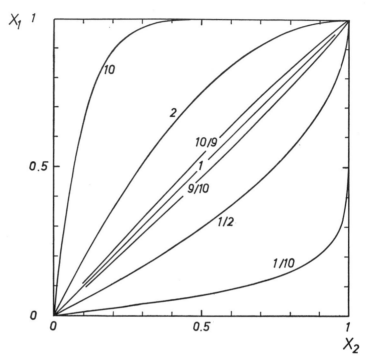

Fig. 3–1. Relation between the amounts of reaction of two competing isotopic molecular species; cf. Eq. (3–3). The figures on the curves indicate the rate-constant ratio k_1/k_2.

$$\frac{\text{remaining fraction of (1)}}{\text{remaining fraction of (2)}} = \frac{1 - x_1}{1 - x_2} = (1 - x_2)^{(k_1/k_2)-1}$$

Except for $k_1/k_2 = 1$, this ratio tends toward zero or infinity when x_2 approaches unity, toward zero for $k_1 > k_2$, and toward infinity for $k_1 < k_2$.   It is frequently claimed that, once a conversion is quantitative, isotope fractionation is unable to cause any change in the isotopic composition.   That is certainly true, but from the reasoning above it is also evident that the claim for completeness of the conversion might be very rigorous if the rates differ considerably, as in most studies with isotopic hydrogen.

TRACER LEVEL
Change of Specific Tracer Content of Reactant

A very common practical case is that in which $A_{(1)}$ occurs only in tracer concentrations in a macroscopic sample of $A_{(2)}$. x_2 could then be treated as the extent of the macroscopic reaction. What is frequently measured is the molar content of the tracer in the remaining reactant, $a_1/(a_1 + a_2) \approx a_1/a_2$. The initial value was a_1^0/a_2^0. The value of the molar content relative to the initial one is consequently

$$\frac{a_1/a_2}{a_1^0/a_2^0} = \frac{a_1/a_1^0}{a_2/a_2^0} = \frac{1 - x_1}{1 - x_2} = (1 - x_2)^{(k_1/k_2)-1} \qquad (3\text{–}4)$$

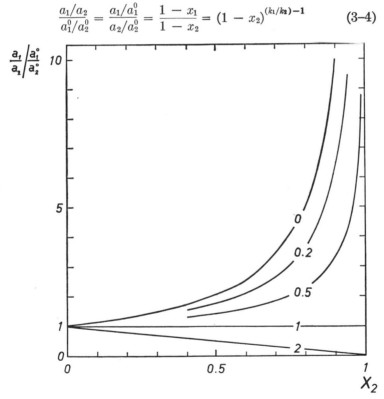

Fig. 3–2. Relation between the relative tracer content of remaining reactant and the amount of reaction; cf. Eq. (3–4). The figures in the curves indicate the rate-constant ratio k_1/k_2.

This equation gives the relation between the relative molar tracer content of the remaining reactant and the amount of reaction; cf. Fig. 3–2. As is evident from the discussion in the preceding section and from Fig. 3–2, the isotopic composition of the last traces of the substrate will be extremely sensitive to the value of k_1/k_2. This fact is worth consideration in the planning of experiments.

The ratio k_1/k_2 might obviously be computed from $a_1 a_2^0/a_1^0 a_2$, and the amount of reaction x_2 by means of Eq. (3–4).

Molar Tracer Content of Products

If, instead, the molar content of the tracer in one of the products is measured and compared to that of the initial reactant, the following relations are valid. The amount of tracer having undergone conversion is $a_1^0 - a_1$, and that of the macroscopic species is $a_2^0 - a_2$. If the fraction p of the tracer molecules is converted in such a way that the tracer atom is contained in the product studied, the ratio r between the molar contents of the product and the initial reactant will be

$$r = \frac{p \dfrac{a_1^0 - a_1}{(a_1^0 - a_1) + (a_2^0 - a_2)}}{\dfrac{a_1^0}{a_1^0 + a_2^0}} \approx p\, \frac{a_1^0 - a_1}{a_1^0} \times \frac{a_2^0}{a_2^0 - a_2} = p\, \frac{x_1}{x_2}$$

$$(3\text{–}5)$$

Hence

$$x_1 = \frac{r}{p} x_2$$

and, according to Eq. (3–2),

$$\frac{k_1}{k_2} = \frac{\log\left(1 - \dfrac{r}{p} x_2\right)}{\log\,(1 - x_2)} \tag{3–6}$$

One Reactive Position (Intermolecular Competition Only). With molecules which have but one way of reacting, only one of the products will contain the tracer. Hence, the molar tracer content of that product is determined by the above equations with $p = 1$.   Therefore k_1/k_2 might be computed directly from r and x_2 by means of Eq. (3–6). r as a function of x_2 is in this case

$$r = \frac{1}{x_2}\left[1 - (1 - x_2)^{k_1/k_2}\right] \qquad (3\text{–}7)$$

This function is shown in Fig. 3–3.

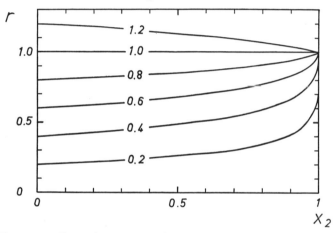

Fig. 3–3. Ratio between molar tracer contents of product and initial reactant as a function of the amount of reaction. Reaction in one single position. Cf. Eq. (3–7). The figures in the curves indicate the rate-constant ratio k_1/k_2.

Several Reactive Positions (Intra- and Intermolecular Competition). The case of a reactant with several chemically equivalent ways of reacting introduces no particular problems if the tracer follows only one of the products.   If, in Eq. (3–6), we simply insert $p = 1$, the isotope effect will appear only in the ratio k_1/k_2. Each is, of course, now a sum of partial constants, which might be differently influenced by the introduction of isotopic atoms.

If the path of the tracer atom is branched, i.e., if the tracer has possibilities of both being and not being transferred in the reaction, two kinds of products will contain it. The branching ratio will itself be a function of the intramolecular isotope effect. In Eq. (3–6), k_1/k_2 as well as p will contain isotope effects, and the equation to be solved after an observation of x_2 and r is transcendental if the intra- and intermolecular isotope effects can be related. Generally they are assumed to have the same strength. A graphical treatment is frequently advantageous, for instance a representation of r versus x_2 for different trial values of the isotope effect, because such a plot shows clearly the influence of the experimental error. The function to be represented is

$$r = \frac{p}{x_2}\left[1-(1-x_2)^{k_1/k_2}\right] \tag{3–8}$$

and p and k_1/k_2 have to be expressed in the isotope effect and numbers given by the symmetry properties of the molecule.

A common case is that in which one of the isotopic molecules, say (2), has the same partial specific rate k' for all n ways of reaction, and the other molecule, (1), has only the speed of one of these alternative ways changed by isotopic substitution. Then

$$\frac{k_1}{k_2} = \frac{(n-1)k' + k''}{nk'} = \frac{n-1+\beta}{n} \tag{3–9}$$

where k'' is the modified partial specific rate and $\beta = k''/k'$ is the isotope effect to be determined.

The isotopic atom is assumed to be transferred in the reaction which corresponds to k''. The probability that the rare isotope will be transferred in the reaction is consequently the fraction

$$p'' = \frac{k''}{(n-1)k' + k''} = \frac{\beta}{n-1+\beta} \tag{3–10}$$

and the probability that it will not be transferred is

$$p' = \frac{(n-1)k'}{(n-1)k' + k''} = \frac{n-1}{n-1+\beta} \tag{3–11}$$

If we choose to study the product into which the reacting atom has been incorporated, the ratio of molar tracer contents will be

$$r'' = \frac{\beta}{n - 1 + \beta} \times \frac{1}{x_2}\left[1 - (1 - x_2)^{(n-1+\beta)/n}\right] \quad (3\text{–}12)$$

At conversions low enough for the square bracket in Eq. (3–12) to be well represented by the linear part of its expansion,

$$r'' = \frac{\beta}{n} \quad (3\text{–}13)$$

Measurement of r'' at low conversions represents a very favorable situation for the determination of β, partly because of the direct proportionality between the two magnitudes (which means a high sensitivity irrespective of the absolute magnitude of β) and partly because the amount of reaction need scarcely be known once it is small. On the whole, this type of measurement is advantageous at any degree of conversion and independently of the number n. The function (3–12) with $n = 2$ is depicted in Fig. 3–4.

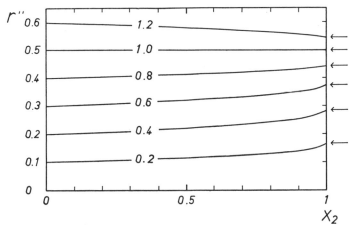

Fig. 3–4. Ratio between molar tracer contents of product containing transferred atom and initial reactant as a function of the amount of reaction; cf. Eq. (3–12), $n = 2$, β as indicated in the curves. Arrows indicate value for $x_2 = 1$.

Sometimes circumstances of the technical kind make the determination of r'' and of the molar tracer content of the recovered reactant less attractive. There remains the possibility of studying the product which contains the atoms which had a chance to react but did not do so. The ratio between the molar tracer content of the latter product and that of the reactant is obtained from Eqs. (3–8), (3–9), and (3–11):

$$r' = \frac{n-1}{n-1+\beta} \times \frac{1}{x_2}\left[1 - (1-x_2)^{(n-1+\beta)/n}\right] \qquad (3\text{–}14)$$

This function has been studied at some length by Melander (64). It is shown for $n = 2$ in Fig. 3–5, and for $n = 6$ in Fig. 3–6.

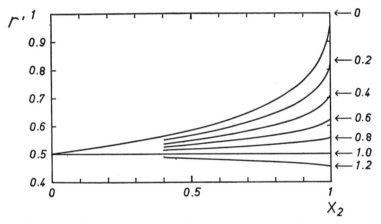

Fig. 3–5. Ratio between molar tracer contents of product containing atoms not transferred and initial reactant as a function of the amount of reaction; cf. Eq. (3–14), $n = 2$, β as indicated at the arrows, which point at value for $x_2 = 1$.

At complete conversion,

$$r'_{x_2=1} = \frac{n-1}{n-1+\beta} \qquad (3\text{–}15)$$

and, at low conversions,

$$r' \approx \frac{n-1}{n} \qquad (3\text{–}16)$$

which means that r' varies in a fairly narrow interval. The sensitivity of r' to variations in β becomes less with increasing number of equivalent positions, n, and with decreasing amount of reaction. In particular, it should be noted that a determination of $r' = (n-1)/n$ within a certain limit of accuracy is of no value unless the amount of reaction simultaneously exceeds a certain corresponding value, because, at a decreasing amount of reaction, r' approaches the value $(n-1)/n$, independently of the isotope effect β. On the whole, the last-mentioned method is the least advantageous one from the purely theoretical standpoint, but it may be the single practicable possibility for experimental reasons.

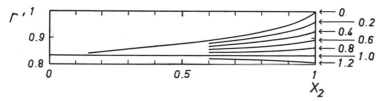

Fig. 3–6. Ratio between molar tracer contents of product containing atoms not transferred and initial reactant as a function of the amount of reaction; cf. Eq. (3–14), $n = 6$, β as indicated at the arrows, which point at value for $x_2 = 1$.

LINEAR APPROXIMATION

It is evident from Eq. (3–3) and Fig. 3–1 that, if the amount of reaction of both isotopic species keeps within a comparatively low value, the curves could be approximated fairly accurately by straight lines. This is the situation when the isotopic material is present in high excess, a situation which is frequently convenient from experimental and computational points of view. The excess should, of course, be estimated primarily with respect to the fast-reacting species, which is most critical. As a matter of fact, most isotopes which are used in tracer amounts are heavier than the ordinary ones and tend to slow down the reactions in which they take part. Consequently the amount of reaction of such a tracer compound will generally not exceed that of the main one, which contributes the bulk of material, and will be the natural basis for the calculation of the proportions of reactants in the batch. It should be pointed out here, however, that the following treatment is not limited to tracer proportions.

For small conversions, i.e., small values of x, Eq. (3–2) might be approximately written

$$\frac{k_1}{k_2} \approx \frac{x_1}{x_2} \qquad (3\text{–}17)$$

This means that the amounts of reaction are proportional to the specific rates. This approximation is equivalent to a replacement of the curves in Fig. 3–1 by their tangents in the origin. It is easy to realize this behavior of the reactions, which are presumed to be of first order with respect to the isotopic species and which do not use up more than a minor fraction of the latter.

Several Equivalent Positions. We shall assume that the isotopic atom occurs in several equivalent positions. If only one of these reacts and if the amount of reaction keeps within the "linear region" as defined above, a very simple treatment is possible when the composition of the product containing the transferred atom is determined. This composition will be constant and will be the same as if each position reacted quite independently, i.e., as if the positions were not grouped together in molecules. Thus, for instance, the hydrogens of water or those of a methyl group could be regarded as independent rather than as occurring in pairs or three and three, respectively.

We shall assume that the intra- and intermolecular isotope effects are equal, and that atom $A_{(1)}$ reacts with specific rate k', and $A_{(2)}$ with specific rate k''. The following scheme applies:

$$(a_1^0) \qquad PA_{(1)m}A_{(2)n} \begin{cases} \xrightarrow{\ mk'\ } & RA_{(1)} \\[1em] \xrightarrow{\ nk''\ } & RA_{(2)} \end{cases}$$

$$(a_2^0) \qquad PA_{(2)m+n} \xrightarrow{\ (m+n)k''\ } RA_{(2)}$$

The initial concentrations are given in parentheses to the left. Application of the approximate Eq. (3–17) gives

$$\frac{mk' + nk''}{(m + n)k''} = \frac{x_1}{x_2} \qquad (3\text{–}18)$$

where the amounts of reaction, x, have been indexed in the same way as the initial concentrations.

We are now to look for the fraction of $RA_{(1)}$ in total RA formed. It is evidently

$$\text{Fraction } RA_{(1)} \text{ in total RA} = \frac{\dfrac{mk'}{mk' + nk''}, x_1 a_1^0}{x_1 a_1^0 + x_2 a_2^0} =$$

$$= \frac{m \dfrac{k'}{k''} a_1^0}{\left(m \dfrac{k'}{k''} + n\right) a_1^0 + (m + n) a_2^0} \tag{3-19}$$

where the last member was obtained by means of Eq. (3–18). As long as the conversion is small enough for the linear approximation to be applicable, the fraction of $RA_{(1)}$ in total RA is obviously independent of the amounts of reaction.

For comparison, the same magnitude will now be calculated, the atoms A being regarded as independent. Using b and y instead of the usual a and x, we obtain

$$(b_1^0) \qquad A_{(1)} \xrightarrow{\ k_1\ } RA_{(1)}$$

$$(b_2^0) \qquad A_{(2)} \xrightarrow{\ k_2\ } RA_{(2)}$$

$$\frac{k_1}{k_2} = \frac{y_1}{y_2} \tag{3-20}$$

$$\text{Fraction } RA_{(1)} \text{ in total RA} = \frac{y_1 b_1^0}{y_1 b_1^0 + y_2 b_2^0} = \frac{\dfrac{k_1}{k_2} b_1^0}{\dfrac{k_1}{k_2} b_1^0 + b_2^0} \tag{3-21}$$

In order to compare Eqs. (3–19) and (3–21) it should be remembered that

$$b_1^0 = m a_1^0$$

$$b_2^0 = n a_1^0 + (m + n) a_2^0$$

thus giving in Eq. (3–21)

$$\text{Fraction RA}_{(1)} \text{ in total RA} = \frac{m \dfrac{k_1}{k_2} a_1^0}{\left(m \dfrac{k_1}{k_2} + n\right) a_1^0 + (m + n)a_2^0}$$

$$(3\text{--}22)$$

Equations (3–19) and (3–22) are identical provided that k'/k'' $= k_1/k_2$. The two treatments consequently give the same value of the isotope effect.

It should be noted that this application of the linear approximation is not limited to tracer proportions between the isotopic species.

ISOTOPE EFFECT IN ISOTOPE EXCHANGE

Owing to their symmetry between reactants and products, isotope exchange reactions are very suitable for the detailed study of mechanism. Hitherto very little has been done in the field of isotope effects in isotope exchange, a field that will certainly prove profitable in the future. From the organic chemist's point of view it is fortunate that hydrogen and carbon have three isotopes each (H^1, H^2, H^3 and C^{12}, C^{13}, C^{14}), long-lived enough to permit convenient handling. One pair of isotopes permits one exchange reaction to be studied, and with three isotopes, in principle, three exchange reactions could be studied. Generally the most convenient way is to let the main isotope form the bulk of material and arrange exchange reactions between it and each of the other two, the latter being present in minute amounts. This method involves the advantage of having both exchange reactions proceed in approximately the same medium even if the solvent is one of the exchange partners.

It is well known that the exchange of isotopes which all react at the same rate (as heavy elements do, approximately) proceeds according to a first-order rate law irrespective of the mechanism. Pertinent references could be found in the monograph of Frost and Pearson (37a). This simple kinetic behavior will still be shown in the presence of an appreciable isotope effect, provided that one of the isotopes is present in very small proportion, as in tracer experiments (25, 46). For comparable amounts of both isotopes, the kinetics will be less simple and will depend on the mechanism and the isotope effect. Since in many studies neither of these is initially known, it might be useful to see what conclusions could be drawn without a knowledge of the mechanism.

From the statements above it is evident that experiments with one of the exchanging isotopes (or isotopic groups) present in very small proportion offer advantages. Present techniques generally make possible the accurate determination of even small fractions of stable isotopes. The following simple discussion is mainly the same as that published in a brief note by Melander (66).

Assume that we have two kinds of molecules, A and B, present in the same homogeneous phase, or each one in its own homogeneous phase. Some kind of atom or group, which occurs in a single kind of position in each molecule and all specimens of which are at first considered to be indistinguishable, will pass from A to B in a molar amount R per unit time. Simultaneously an equal amount will pass from B to A. Since the exchanged and non-exchanged specimens are undistinguishable, we cannot determine R. A and B are assumed to be present in macroscopic amounts, and the molar amount of the exchangeable species in each is denoted by a and b, respectively. Nothing in the macroscopic behavior is changed if we replace a very minute amount of the exchangeable entity by a distinguishable one. At one instant the fractions in A and B are y and z, respectively. Were there no isotope discrimination, the amount of distinguishable entity transferred from A to B in unit time would be yR, and that in the opposite direction would be zR. This is merely a matter of statistics.

If there is an isotope effect, for one reason or another, the chance for the tracer species to be transferred in one direction is different from that for the ordinary atoms or groups, and this must be accounted for by the probability factors α and β, the rates being now $y\alpha R$ and $z\beta R$. α and β are the kinetic isotope effects in the two opposite transfer reactions. At exchange equilibrium there is no net transport, making

$$y_\infty \alpha R = z_\infty \beta R$$

or

$$\frac{z_\infty}{y_\infty} = \frac{\alpha}{\beta} \qquad (3\text{--}23)$$

The ratio α/β is consequently the isotope exchange equilibrium constant.

If there is no equilibrium as yet, the net transport rate from A to B will be

$$-\frac{d(ya)}{dt} = y\alpha R - z\beta R \qquad (3\text{--}24)$$

and, if the fractions were y_0 and z_0 at zero time, the conservation of matter gives

$$ya + zb = y_0 a + z_0 b = y_\infty a + z_\infty b \qquad (3\text{--}25)$$

It is convenient to introduce a new variable, the fractional exchange F, defined as

$$F = \frac{y - y_0}{y_\infty - y_0} = \frac{z - z_0}{z_\infty - z_0} \qquad (3\text{--}26)$$

[That the two ratios are equal follows from Eq. (3–25).] If Eqs. (3–23), (3–25), and (3–26) are introduced into Eq. (3–24), the latter acquires a very simple form which is integrated:

$$\int_0^F \frac{dF}{1 - F} = \frac{\beta a + \alpha b}{ab} R \int_0^t dt$$

giving

$$-\ln(1 - F) = \frac{\beta a + \alpha b}{ab} Rt \qquad (3\text{--}27)$$

If the reaction can be carried on until equilibrium is established, y_∞ and (or) z_∞ might be determined, and the fractional exchange could be calculated from Eq. (3–26). When $-\ln(1 - F)$ is plotted versus time, a straight line of slope $(\beta a + \alpha b)R/ab$ is obtained. By means of this slope and the ratio α/β from Eq. (3–23) it is possible to calculate αR and

βR, but these two magnitudes cannot be separated into their factors. Thus the absolute isotope effects cannot be computed; neither can the fundamental exchange rate R.

When another tracer isotope is available, the experiment can be repeated under the same conditions. Then the rate R is, of course, the same, and two new values, $\alpha' R$ and $\beta' R$, are obtained. From the two experiments, α'/α and β'/β and hence the influence of the isotopic mass on the exchange rate are determined.

Initial Rate Studies. In several experiments it is impossible to use the accurate and convenient method described above because side reactions cause too much disturbance before equilibrium is established. Under such circumstances, however, it is frequently possible to study the initial rate. The net transport rate could be written

$$-\frac{d(ya)}{dt} = \text{rate}_{A \to B} - \text{rate}_{B \to A}$$

where "rate" refers to the transport of the distinguishable entity; y is, as usual, the fraction of it in A. If B contains none of this isotope initially, we obtain

$$- a \left[\frac{dy}{dt}\right]_{t=0} = \left[\text{rate}_{A \to B}\right]_{t=0} \tag{3-28}$$

For two similar studies with different distinguishable isotopes, the ratio of the initial decay rates of y will obviously give the exchange isotope effect. If the amount b is chosen considerably larger than a, the increase in z will be slow and the back reaction will be without appreciable influence for a period long enough for an accurate determination of $[dy/dt]_{t=0}$. This determination, however, generally requires some care and constitutes the drawback of the method.

It should be observed that the latter method does not require the fraction y to be infinitely small. With considerable amounts of the distinguishable isotope, however, it is necessary to observe that mechanisms which offer the possibility for more than one distinguishable molecule to enter the same transition state could make the isotope effect less significant. With tracer amounts the chance of such coincidences is entirely negligible.

With the formalism of Eq. (3–24), valid at tracer levels when there exists a fundamental rate R, the last-mentioned method gives

$$- a \left[\frac{dy}{dt} \right]_{t=0} = y_0 \alpha R$$

or

$$- \left[\frac{d \ln y}{dt} \right]_{t=0} = \frac{\alpha R}{a} \tag{3–29}$$

As with the first method, αR is determined. In principle, it would be possible to determine βR in the same way by arranging exchange experiments in the opposite direction. This would lead to an indirect determination of the exchange equilibrium constant α/β. It should be observed, however, that R is dependent on the amounts of a and b, and, if the mechanism is unknown, the dependence of R is also unknown. It would thus be necessary to carry out both exchange reactions, A$\longrightarrow$B and B$\longrightarrow$A, with the same choice of a and b, and this proportion could not be chosen favorably for the determination of the initial rate of both reactions. On the other hand, it is still possible to carry out an accurate determination in the same direction with constant R for each of two distinguishable isotopes, giving α'/α. With other proportions of a and b and hence another R, it is also possible to determine β'/β. Thus the method still allows the determination of the kinetic isotope effect for the two opposite reactions.

4

Primary
Hydrogen Isotope Effect
in Single Reaction Steps

As seen in Chapter 2, the theoretical computation of kinetic isotope effects is founded on the theory of absolute reaction rates, which in itself is founded on certain postulates. Even if it were perfectly reliable in all cases, the computation requires such detailed information about the transition state that we can never hope to know enough about the latter for a precise calculation in most cases of practical value to the chemist. Hence it is of primary importance to investigate as simple reactions as possible in order to test the predictions and in order to obtain an empirical material which will be useful, for instance, when the measurement of kinetic isotope effects is to be used as a diagnostic tool to unravel complicated reaction mechanisms.

That our assumptions are mainly correct is evident from studies of very simple reactions in the gas phase, for which fairly accurate calculations are still possible. Thus Johnston (54) has recently reported briefly that he has successfully treated the reaction between a methyl radical and a hydrogen molecule:

$$x_3C + yz \longrightarrow x_3C \text{----} y \text{----} z \longrightarrow x_3Cy + z$$
(transition state)

where x, y, and z might be protium or deuterium. A complete vibrational analysis was possible for the transition state, and the force constants were obtained from considerations of molecular structure and spectroscopy. The frequencies of the eight isotopic transition states could thus be found, as well as the reduced mass along the reaction coordinate. The isotope effect was then calculated by means of Bigeleisen's expression, and eighteen different rate-constant ratios were in satisfactory agreement with experimental results of Whittle and Steacie.

Gas Phase Decomposition of Chloroform. It might be of interest to examine one case of decomposition in the gas phase. Dibeler and Bernstein (31) have studied the hydrogen isotope effect on the dissociation probability of chloroform in the mass spectrograph. There might be some doubt whether there is justification for treating this reaction (occurring in the ion source) as an ordinary thermal (high-pressure) decomposition reaction, and the results obtained below should perhaps be considered more or less as proof that it is justified.

Without entering the physical details, Dibeler and Bernstein found that deuterium in CCl_3D was split off 0.33 times as frequently as protium in CCl_3H, i.e., $k_D/k_H = 0.33$, when the temperature of the ion source was 245°C.

The vibrational frequencies required for a rough calculation of the isotope effect could be obtained from a paper by Wood and Rank (104). A vibration which could approximately be assigned to the stretching of the carbon–hydrogen bond has the frequency 3019 cm^{-1} in CCl_3H, and 2256 cm^{-1} in CCl_3D. The twofold degenerate bending frequency of the same bond could be assigned the values 1216 and 908 cm^{-1}, respectively, for the two molecules. The dissociation of chloroform will now be treated according to the case of "isotopic hydrogen in heavy molecules," which ought to be appropriate.

If the bending frequencies are assumed to be mainly unchanged in the transition state, Eq. (2–10) should be used. Since the transition state certainly has the same symmetry as the initial molecule, the symmetry numbers are cancelled. We obtain

$$\frac{k_D}{k_H} = \exp\left\{\frac{hc}{2k \times 518}(2256 - 3019)\right\} = 0.35$$

This result is in good agreement with the observed value, and the computation is practically equivalent to that of Eyring and Cagle (35), who used Eq. (2–7).

In the present reaction, however, it does not seem quite unlikely that the bond to be broken is stretched very much in the transition state and, consequently, that the forces opposing the bending are weakened considerably. Such a case might be better represented by the situation corresponding to Eq. (2–12), which gives

$$\frac{k_D}{k_H} = \frac{2}{1}\exp\left\{\frac{hc}{2k \times 518}\left[2256-3019+2(908-1216)\right]\right\} = 0.29$$

Thus both pictures give about the same prediction regarding the isotope effect. Both values also agree well with the experimental one.

It might perhaps seem astonishing that the result is not appreciably affected by the large difference in assumptions made in this case. The agreement is fortuitous, however, and with the same set of frequencies it would not hold at other temperatures.

HYDROGEN ABSTRACTION IN GENERAL

Except for decompositions at high temperatures, most simple reactions involving the splitting off of hydrogen from a residue of some kind could properly be termed hydrogen "abstraction" reactions, a third body accepting the hydrogen always being present. Reactions of this kind play an important role in organic chemistry. The acceptor is sometimes of radical character, as in the examples to be discussed in the next section. A very important case is the nucleophilic attack on hydrogen by bases of different kinds, particularly the hydroxide ion. In these two cases the hydrogen leaves the residue, taking one or none of its binding electrons along. The third possibility, that the hydrogen takes both electrons of the initial bond with itself, has probably been observed in some cases.

It should be noted that the kinetic isotope effect is not particularly dependent on whether the attack on hydrogen is nucleophilic, radical, or electrophilic. The theoretical expression for the isotope effect does not explicitly account for the behavior of the electrons in this respect; it is only through the potential energy of the pertinent atomic conformations, which certainly is governed mainly by the electrons, that the electrons influence the isotope effect. If the transition state is not very similar to the reaction product, its potential energy and the variation of this energy with minor vibrational movements are not particularly sensitive to the ultimate destiny of the electrons.

HYDROGEN ABSTRACTION FROM THE SIDE CHAIN OF TOLUENE BY RADICAL ATTACK

Chlorine Atoms. First we are to consider the reaction between the methyl hydrogens of toluene and chlorine atoms, generated from chlorine molecules by light. The work in this field is still in progress, and the results of different authors do not quite agree. The reaction is chosen, however, as the different investigators have used very different isotope techniques the study of which might be instructive.

The step which is actually studied in this reaction is the abstraction of methyl hydrogens by chlorine atoms. It seems fairly safe to assume that all benzyl radicals formed will join with chlorine atoms without any intervening hydrogen exchange reactions, and consequently the isotopic composition of the unreacted material, the benzyl chloride and the hydrogen chloride, will be representative of the primary hydrogen abstraction.

With the two isotopic molecules $C_6H_5CH_2D$ and $C_6H_5CH_3$, the reaction scheme would be as follows, provided that the intra- and intermolecular isotope effects are assumed to be equal:

$$C_6H_5CH_2D + Cl \cdot \begin{cases} \xrightarrow{k_D} & C_6H_5CH_2 \cdot \; + DCl \\ \xrightarrow{2k_H} & C_6H_5CHD \cdot + HCl \end{cases}$$

$$C_6H_5CH_3 + Cl \cdot \quad \xrightarrow{3k_H} \quad C_6H_5CH_2 \cdot \; + HCl$$

The numbers in the specific rates are easily found intuitively and may also be found from symmetry considerations as described below. Since this is a case of isotopic hydrogen in a heavy molecule, the approximate Eq. (2–10) or Eq. (2–12) could be applied.

The methyl carbon might be regarded as the center of the molecule. As the internal symmetry number for the rotation of the phenyl group is the same for any of the transition states as for the initial hydrocarbon, it could be omitted. If the reactant is ordinary toluene, the methyl group will introduce the symmetry number 3, while the monodeuterated methyl group of the competing reactant will introduce the corresponding number 1.

In the transition state the carbon-hydrogen bond concerned is probably stretched and the chlorine atom is probably situated on the straight line through these atoms outside the hydrogen. If all methyl hydrogens are protium atoms, the symmetry number introduced by the reacting methyl group will be 1. If one of the methyl hydrogens not taking part in the reaction is a deuteron, the transition state will be one of two possible enantiomorphs, the carbon atom being asymmetric. The left-handed and the right-handed transition states contain a stretched carbon-protium bond, which is going to be ruptured. With our present degree of approximation the frequency or frequencies which are lost in passing into the transition state are the same in this case for the two initial molecules considered, the mass of neighboring hydrogen atoms being assumed to have negligible influence on the carbon-protium vibration. For the two enantiomorphs compared to the completely protiated group, the right-hand side of Eq. (2–10) and of Eq. (2–12) is equal to 1. If, finally, the deuteron takes part in the reaction, the symmetry number of the transition state is 1, and now the heavy molecule has lost one or more of the frequencies associated with the carbon-deuterium bond. Thus the frequencies $\nu_{(1)}$ and $\nu_{(2)}$ will no longer match in the right-hand side of Eq. (2–10) or Eq. (2–12), its value being denoted by k_D/k_H. In writing down the rate-constant ratio, k_1/k_2, of the deuterated and the non-deuterated molecules, all three possibilities for the transition state from the deuterated methyl group, the two enantiomorphs with reacting protium, and the single form with reacting deuterium should be accounted for. We obtain

$$\frac{k_1}{k_2} = \frac{1}{3} \times \frac{1}{1} \times 1 + \frac{1}{3} \times \frac{1}{1} \times 1 + \frac{1}{3} \times \frac{1}{1} \times \frac{k_D}{k_H} = \frac{k_D + 2k_H}{3k_H}$$

Using the symbolism of Chapter 3 and denoting $C_6H_5CH_2D$ by index 1 and $C_6H_5CH_3$ by index 2 as above, we obtain, according to Eqs. (3–2) and (3–17),

$$\frac{k_1}{k_2} = \frac{k_D + 2k_H}{3k_H} = \frac{\log(1 - x_1)}{\log(1 - x_2)} \approx \frac{x_1}{x_2}$$

the last approximation being valid for small conversions, i.e., within the "linear region" of Fig. 3–1.

As an example, let us consider the investigation of Brown and Russell (19) on the photochlorination of toluene-α-d in its own liquid phase at 80°C. In one of their experiments they chlorinated 1.46 moles of toluene-α-d, containing 1.6% of ordinary toluene, with 0.1 mole of chlorine. Thus the amount of reaction could not exceed 0.07, and the system is well represented by the linear approximation. According to pp. 58 ff., the fraction of deuterium chloride to be found in the total hydrogen chloride is obtained from Eq. (3–19):

$$\text{Fraction DCl in total hydrogen chloride} = \frac{\dfrac{k_D}{k_H} a_1^0}{\left(\dfrac{k_D}{k_H} + 2\right) a_1^0 + 3a_2^0}$$

where a_1^0 and a_2^0 denote the initial concentrations of the two isotopic molecules. We know

$$a_2^0 = 0.016(a_1^0 + a_2^0)$$

from which

$$\frac{a_2^0}{a_1^0} = \frac{0.016}{0.984}$$

The reaction produced 0.0212 mole of deuterium chloride together with 0.0868 mole of hydrogen chloride, whence

$$\frac{\dfrac{k_D}{k_H}}{\dfrac{k_D}{k_H} + 2 + 3 \times \dfrac{0.016}{0.984}} = \frac{0.0212}{0.0212 + 0.0868}$$

or

$$\frac{k_D}{k_H} = 0.50$$

It is of some interest to see what result would have been obtained if we had neglected the 1.6% of ordinary toluene and regarded the sample as pure monodeuterotoluene. The competition would then have been regarded as an exclusively intramolecular one (p. 47). The simple equation would have been

$$\frac{k_D}{2k_H} = \frac{0.0212}{0.0868}$$

or

$$\frac{k_D}{k_H} = 0.49$$

Thus the presence of the light compound would cause only a minor error if the system were treated in this very simple way.

According to studies of the reproducibility of this type of experiment and the competitive chlorination of monodeuterotoluene and ordinary cyclohexane, the possibility of appreciable hydrogen abstraction by hydrocarbon radicals during the photochlorination could be excluded. Within the experimental accuracy there was agreement between the amounts of benzyl chloride and chlorocyclohexane formed and the proportions of deuterium and protium in the hydrogen chloride evolved. For the calculation the results obtained as above were used.

Similar competition experiments between monodeutero-toluene and undeuterated (but in different other ways substituted) toluenes have been carried out by Walling and Miller (98) in order to study the relative reactivities of substituted toluenes toward chlorine atoms. Obviously, in order to relate the reactivities to that of undeuterated toluene, the isotope effect in the chlorination of toluene without further substituents must be known, and the investigation includes determinations of this effect at 69.5°C in various solvents and in the gas phase.

Walling and Miller used samples with monodeuterotoluene in ordinary toluene, the heavy compound being the minor component. Therefore the reaction could not be treated, even approximately, in the simpler of the two ways above. In order to avoid multiple chlorination, almost tenfold excess of toluene was used. In the kinetic calculations all pertinent hydrogens were treated as independent, their grouping together in molecules being neglected (pp. 58 ff.). This is permissible since the proportions of reactants warranted the applicability of the "linear approximation."

The approximate reaction scheme is consequently that of p. 59:

$$-D + Cl\cdot \xrightarrow{k_D} DCl$$

$$-H + Cl\cdot \xrightarrow{k_H} HCl$$

and Eq. (3–20) gives

$$\frac{k_D}{k_H} = \frac{x_D}{x_H}$$

In one experiment 0.271 mole of the mixture of the isotopic toluenes and 0.041 mole of chlorine were used. Combustion of a sample of the initial toluene gave water containing 2.555% of deuterium.* Since all this deuterium was located in the methyl group, maximum one atom in each molecule, the initial

* According to a personal communication from Dr. C. Walling, this figure should replace the one erroneously given in the paper referred to.

amount of $C_6H_5CH_2D$ should have been $8 \times 0.02555 \times 0.271$ $= 0.0554$ mole, and thus that of $C_6H_5CH_3 = 0.215_6$ mole. The hydrogen chloride formed contained 3.45% of deuterium. Hence $0.0345 \times 0.041 = 0.0014_1$ mole of the chlorine had reacted with deuterium, and the other 0.039_6 mole with protium. The rate-constant ratio is

$$\frac{k_D}{k_H} = \frac{\dfrac{0.00141}{0.0554}}{\dfrac{0.0396}{2 \times 0.0554 + 3 \times 0.2156}} = 0.49$$

In a recent investigation Wiberg and Slaugh (102) studied the same chlorination in a different way. They used, for instance, a mixture of ordinary toluene and monodeuterotoluene the molecular composition of which was determined by mass spectrometric analysis. A fairly large known fraction was converted to benzyl chloride, and determinations of the composition of the recovered reactant as well as that of the produced derivative were carried out afterwards. In this way it was possible to determine to what extent the converted $C_6H_5CH_2D$ had reacted with either of its two kinds of hydrogen. This gives directly the branching ratio and the intramolecular isotope effect. For the chlorination with chlorine in carbon tetrachloride solution at 77°C the intramolecular isotope effect was $k_D/k_H = 0.77$ for $C_6H_5CH_2D$, and 0.72 for $C_6H_5CHD_2$, the uncertainty in the latter figure being just sufficient for a possible agreement with the first one. On the other hand, the results, in general, indicate that the intramolecular isotope effect is not quite independent of the mass of neighboring hydrogens but becomes slightly stronger with further deuteration. The reason for this is not very well known. It should be observed that such subtle effects are not taken into account by our simple models, in which each bond is assumed to react with specific rate k_D or k_H, irrespective of the mass of the neighbors.

The k_D/k_H values for the photochlorination of toluene obtained by investigators who used different techniques differ appreciably. The reason for this conflicting evidence is not known at present. Walling and Miller (98) claim values within the interval 0.48 ± 0.01 at 69.5°C for the reaction in the vapor phase, in its own liquid phase, and in the solvents benzene and o-dichlorobenzene. In carbon tetrachloride, in which solvent the behavior is somewhat different in other respects too, they obtained the value 0.503. It is interesting to note that with sulfuryl chloride in liquid toluene the value was 0.475, suggesting attack by chlorine atoms as the main step for this reaction also. Brown and Russell (19) obtained the value 0.50 with chlorine and light in liquid toluene phase at 80°C. Wiberg and Slaugh (102) obtained 0.72 and 0.77 (cf. above) in carbon tetrachloride at 77°C, and 0.68 in liquid toluene at 110°C. Under the same conditions sulfuryl chloride gave 0.70 and 0.58, respectively. These results have recently been further compared and discussed by Russell (85).

Bromine Atoms. All isotope effects reported for the chlorination of toluene could certainly be considered very weak in view of the fact that a carbon-hydrogen bond is opened in the reaction. Things are different with photobromination, which is otherwise analogous to photochlorination. Thus Wiberg and Slaugh (102) give values of $k_D/k_H = 0.19$ to 0.22 (approximately) for the bromination of isotopic toluenes in carbon tetrachloride at 77°C.

Organic Radicals. The methyl hydrogen might also be abstracted by radicals other than halogen atoms. Thus the decomposition of peroxides in toluene or p-xylene gives radicals which abstract hydrogen from the methyls, the reaction leading to dimerization of the generated benzyl or p-methylbenzyl. Eliel et al. (34, 103) have studied this reaction and found isotope effects of the order of $k_D/k_H = 0.1$ to 0.2 for toluene and xylene in (mainly their own) liquid phase at about 120°C.

Discussion. The isotope effects in hydrogen abstraction from the methyl group in toluene cover almost the entire interval between a very weak effect and the theoretically strongest possible effect, as is evident from a comparison with Table 2–1, p. 22. It is interesting to find that the activation energies increase in the same sequence as that in which the isotope effect becomes stronger. In most three-center reactions the stretching of the bond to be finally broken makes a considerable contribution to the activation energy. It is to be expected that the reaction with chlorine, which has a very small activation energy and is strongly exothermic, reaches its transition state fairly early along the reaction coordinate, i.e., before the carbon-hydrogen bond has been stretched very much. The transition state will be fairly similar to the initial toluene plus the chlorine atom, somewhat like X + YZ in Fig. 2–3. The initial bond keeps a good deal of its zero-point energy, and the presence of the chlorine might even increase the bending frequencies, thus making the isotope effect weaker than predicted by Eq. (2–10). The reaction with bromine atoms is less exothermic and also has a higher activation energy. It is to be expected that the fission of the carbon-hydrogen bond will be more advanced in this case, thus also rendering weaker the forces opposing bending. Figure 2–4 is probably a better approximation than Fig. 2–3 for this case. The reaction with organic radicals, finally, represents a case still more advanced in the same direction, although such a reaction is generally fairly exothermic. To judge from the experimental and calculated isotope effects, it could be expected that the frequencies of bending vibrations should be very low, Eq. (2–12) being a better approximation than Eq. (2–10). For the energetic relations concerning these reactions the reader is referred to the above investigations and the recent monograph by Walling (97). The general relation between the state in the activated complex and the degree of advancement along the reaction coordinate has been discussed by Hammond (45), and this particular situation has been discussed by Wiberg (101).

It has frequently been claimed that there should be a general relation between rate and isotope effect within a group of analogous reactions, high activation energies being accompanied by strong isotope effects. Since what is meant here by "analogous" reactions will generally involve reactions with mainly constant entropy of activation, the correlation could also be said to be one of strong isotope effect with low rate. The reasons should be obvious from the discussion above. Thus, substituted toluenes have been found by Wiberg and Slaugh (102) to have isotope effects which become stronger with decreasing reaction rate. The compounds p-$CH_3O \cdot C_6H_4 \cdot CH_2D$, $C_6H_5CH_2D$, and p-$ClC_6H_4 \cdot CH_2D$, which are mentioned in the order of decreasing rate, or increasing activation energy, in their reaction with N-bromosuccinimide at 77°C, give the intramolecular isotope effects k_D/k_H = 0.31, 0.21, and 0.20, respectively. The reaction probably involves an attack by the N-succinimidyl radical on the hydrogen.

This empirical rule, however, should be handled with care in cases like the last-mentioned one, in which the isotopic atom initially is situated in different molecules. Thus the potential-energy curve for the stretching of the carbon-hydrogen bond might have different shapes for different derivatives, particularly in regions of appreciable stretching, for which the force cannot be evaluated from ordinary infrared absorption data. Consequently it cannot be entirely safe to assume some unequivocal relation between stretching and activation energy, even if the zero-point energies are about the same in the different initial molecules.

CARBANION FORMATION

Formation of a carbanion through abstraction of a hydrogen by a nucleophilic reagent is an important and rate-determining step in many reactions. The reaction occurs with substances containing C—H bonds with certain neighboring groups like the keto or the nitro group. The general electronic mechanism could be pictured where A denotes a carbon or nitrogen atom. Except for the base B, which is necessary with both types of molecules, the reaction with ketones is catalyzed by acids through proton donation to the oxygen. The latter does not apply to nitro compounds, in which the atom A is attached to two oxygens.

The carbanion might undergo different rapid subsequent reactions the observation of which allows a kinetic study of the rate-determining step. Thus bromine (cation) adds very rapidly to the carbanion, a reaction which has been used by Reitz for kinetic studies of the process. The fact that the observed rate is independent of the bromine concentration shows that it is the hydrogen abstraction which is being measured.

Reitz used ordinary acetone and nitromethane and compared their rates of bromination with those of almost completely deuterated samples of the same compounds in independent experiments. By means of extrapolation he determined the reaction rates of the completely deuterated molecules, thus being able to compute the ratio of the specific rates for the reactions of CD_3COCD_3 and CH_3COCH_3, or CD_3NO_2 and CH_3NO_2, respectively.

Nitromethane, brominated at 70°C in ordinary water without the presence of a catalyst, gave $k_D/k_H = 0.26$; in heavy water, 0.19. With monochloroacetate ion as catalyst the corresponding values were 0.23 and 0.19, respectively, at 25°C; and with acetate ion at the same temperature, 0.15 and 0.14 (76). The basicity of the hydrogen abstractor increases in the series water < monochloroacetate ion < acetate ion, and so does the absolute rate of the reaction. As far as the two latter catalysts are concerned, the isotope effect seems to become stronger in the same sequence, but, if the values of the pure water experiments are corrected for temperature, they hardly fall in the series. It is doubtful, however, whether the accuracy permits more exact comparisons.

With acetone, the reaction of which is catalyzed by acids as well as by bases, the isotope effect seems fairly constant under different conditions. Reitz found $k_D/k_H = 0.13$ at 25°C for the reaction catalyzed by strong acids, the value being insensitive to variations of the isotopic composition of the aqueous medium (77). Approximately the same value is obtained also when weak acids or acetate ion is used as catalyst (78).

The role of the different catalysts in this type of reaction is still being discussed, and pure kinetics cannot give unambiguous information. Thus, for instance, proportionality of the rate to acetic acid concentration may mean that acetic acid is operative as the acid and water as the base, but also that the acid may be hydronium ion and the base acetate ion. Any mixture of these mechanisms gives a kinetic expression of the same form. Swain *et al.* (95) have recently discussed this complex problem and made an effort to decide between different possibilities by means of isotope effect studies. Very great care is needed in deciding between different closely analogous transition states from small differences in the isotope effect, even if the error in the measurements is negligible. The isotope effect k_T/k_H in aqueous medium containing dioxane for α-phenylisocaprophenone, $C_6H_5COCH(C_6H_5)CH_2CH(CH_3)_2$, at 98°C was within the range 0.078 to 0.103 for the catalysts hydronium ion, acetic acid, acetate ion, and hydroxide ion.

The investigation of Swain *et al.* is interesting from the experimental point of view. The rate of the hydrogen abstraction from the ketone was measured with quite independent methods for the different isotopes. The loss of protium was followed via the racemization of the optically active ketone, and the loss of tritium (to the solvent) by means of its radioactivity. The optical activity at the center of reaction offers a very nice tool for studying the formation of the carbanion, in which the asymmetry is lost. Since the carbanion is very soon joined again with a hydrogen cation if other reactions do not occur in the system, the loss of optical activity could be said to offer a means of following the replacement of protium by protium, a reaction which could not be detected by other means and hence not with simple compounds devoid of asymmetric centers.

Swain *et al.* also checked the relation between tritium and deuterium isotope effects (Eq. 2–13) by following the loss of deuterium from the same ketone to the solvent. The two isotope effects to be compared were $k_T/k_H = 0.098$ and $k_D/k_H = 0.199$, within the accuracy satisfying the interrelation.

The observed isotope effects in the formation of carbanions by base attack hitherto described are of a fairly uniform strength, not weak but certainly weaker than the limiting strength. That considerably weaker isotope effects could occur also in similar reactions is evident from the work of Hine *et al.* (49), who studied the hydrogen abstraction from haloforms by hydroxide ion. They found, for the reaction

$$CHCl_2F + OH^- \longrightarrow CCl_2F^- + H_2O$$

in aqueous solution at 0°C, $k_D/k_H = 0.57$ and, at 20.2°C, 0.66. For the corresponding reaction,

$$CHBrClF + OH^- \longrightarrow CBrClF^- + H_2O$$

$k_D/k_H = 0.57$ at 0°C, and 0.59 at 15°C.

HYDROGEN CARRIED AWAY WITH BOTH BONDING ELECTRONS

Although the alkaline cleavage of triorganosilanes could as well be treated in Chapter 6 as an example of the investigation of reaction mechanisms by isotopic means, it will be discussed here as an example of a reaction in which hydrogen leaves a heavy residue and carries both bonding electrons away. This type of reaction is not very common, but it represents the third and last possibility regarding the electronic mechanism of hydrogen abstraction. In this case the hydrogen is not abstracted from an otherwise undisturbed center, but the transition state has a termolecular composition. Of course, such complicated reactions are still more difficult to treat from the theoretical point of view than those hitherto dealt with, but nevertheless the reaction offers many interesting features. Moreover, it has been subject to several accurate measurements, the results obtained by different schools being in agreement.

When, for instance, triphenylsilane is treated with a solution containing some alkaline agent and a hydroxyl compound, for instance piperidine with small amounts of water or potassium hydroxide in slightly aqueous ethanol, the following gross reaction takes place:

$$HO^- + (C_6H_5)_3SiH + HOR \longrightarrow (C_6H_5)_3SiOH + H_2 + OR^-$$

where R might be hydrogen or the ethyl residue. Kinetic investigations (see ref. 58 for further references) have shown that the above components are probably required for the formation of the transition state. The weakest kinetic evidence is that for the presence of the solvent molecule HOR. The fact that the isotope effect measured with isotopic silanes is fairly sensitive to the nature of R, however, indicates that the molecule HOR is likely to be involved (58). On the other hand, as will be seen below, this isotope effect is so weak that it might as well have its origin in a slight weakening of the Si—H bond in a simple attack by the hydroxide ion on the silane. The hydrogen could subsequently be carried away by the solvent in a rapid process without influencing the gross rate, i.e., without causing any discrimination between the isotopic silanes. This is not the place, however, to deal with the problem of isotope effects in kinetically complex reactions; this subject is treated in Chapter 6. To sum up, most experimental facts favor the termolecular picture, and some simple models of this transition state will be theoretically treated in the following and the predictions compared with the experimental results.

The transition state has been formulated (38, 58) quite generally in the following way:

$$
\left[
\begin{array}{c}
OH \\
\vdots \\
(C_6H_5)_3Si \text{----} H \text{----} H \text{----} OR
\end{array}
\right]^-
$$

Kaplan and Wilzbach (58) have made calculations on the strongly simplified models (I) and (II) below, using Eq. (2–9) and assuming that all frequencies but those of the stretching modes of the bonds ending at isotopic atoms are cancelled. The "non-isotopic" bonds are likely to give cancellation within each sum of the exponent because $u_{(1)} = u_{(2)}$ and $u_{(1)}{}^{\ddagger} = u_{(2)}{}^{\ddagger}$, and, for example, the bending vibrations could be assumed to be cancelled between the sums because $u^{\ddagger} \approx u$ for each isotope and each similar vibration in the transition state and in the reactants. Of course, as always, it is a rough approximation to talk of molecular vibrations in terms of bond vibrations.

$$\left[\begin{array}{cccc} \text{OH} & & & \\ | & & & \\ | & & & \\ (C_6H_5)_3Si & H & H & OR \end{array} \right]^{-} \qquad (I)$$

$$\left[\begin{array}{cccc} \text{OH} & & & \\ | & & & \\ | & & & \\ (C_6H_5)_3Si & H \text{——} H & OR \end{array} \right]^{-} \qquad (II)$$

In model (I) the hydrogen atoms are assumed to be entirely free, and in model (II) they are assumed to have formed a normal hydrogen-hydrogen bond. Since neither silicon nor the hydroxide oxygen is isotopic in the experiments to be described below, it is not necessary to make up one's mind as to the state of the silicon-oxygen bond. Its vibrational frequencies are assumed to be cancelled in the computations. Per se, it is possible for it to be an ordinary bond, formed in a possible rapidly established pre-equilibrium, because silicon is able to accommodate more than four electron pairs in its valence shell. The following frequencies were assigned to the initial bonds: SiH, 2135; SiD, 1547; SiT, 1282; OH, 3259; and OT, 1988 cm^{-1}. The frequencies of the isotopic hydrogen molecules are: H_2, 4405; HD, 3817; HT, 3598; and DT, 2846 cm^{-1}. (Most of these values are observed; cf. ref. 58 for references.)

If now model (I) is used for the calculation of the isotope effect at 25°C between tritiated and ordinary silane, the expression becomes

$$\frac{k_{\text{SiT}}}{k_{\text{SiH}}} = \exp\left[- \frac{hc}{2k \times 298}(-1282 + 2135)\right] = 0.13$$

because the single "isotopic" bond concerned is that between silicon and hydrogen in the reactant. In the same way we find $k_{\text{SiD}}/k_{\text{SiH}} = 0.24$ at the same temperature.

The measured isotope effect (see below) is much weaker; and this is a reason for trying model (II), since the hydrogen molecule has a particularly high frequency and an increase of the zero-point energy in the transition state will weaken the isotope effect provided this zero-point energy depends on the pertinent isotopic mass. Model (II) with tritiated and ordinary silane gives at 25°C

$$\frac{k_{\text{SiT}}}{k_{\text{SiH}}} = \exp\left[- \frac{hc}{2k \times 298}(3598 - 4405 - 1282 + 2135)\right] =$$
$$= 0.89$$

and similarly $k_{\text{SiD}}/k_{\text{SiH}} = 1.00$, since in the latter case the frequencies happen to be cancelled perfectly.

Most of the corresponding experiments were carried out as competitive reactions, and the evolved hydrogen was quantitatively studied. The hydrogen was collected in fractions, and the isotopic composition of such a fraction told how much of each isotopic triphenylsilane had reacted during the corresponding interval. Equation (3–1),

$$\frac{k_1}{k_2} = \frac{\log (a_1/a_1^0)}{\log (a_2/a_2^0)}$$

could now be used, a_1^0 and a_2^0 in this case denoting the concentrations at the beginning of the interval and a_1 and a_2 those at the end. Since in the present case there is only one reactive position, k_1 and k_2 are the specific rates of direct interest.

Thus, for instance, a certain mixture of ordinary and deuterated triphenylsilane ultimately evolved 0.900_3 mmole of

HD and 0.973_2 mmole of H_2. At one moment it had evolved 0.180_3 mmole of HD and 0.221_6 mmole of H_2. The next fraction consisted of 0.095_9 and 0.114_1 mmoles, respectively, of these compounds. From these figures we obtain

$$\frac{k_{SiD}}{k_{SiH}} =$$

$$= \frac{\log (0.9003 - 0.1803 - 0.0959) - \log (0.9003 - 0.1803)}{\log (0.9732 - 0.2216 - 0.1141) - \log (0.9732 - 0.2216)} =$$

$$= 0.868$$

The corresponding calculation was repeated for each fraction. The mean results obtained by Kaplan and Wilzbach (58) with triphenylsilane and piperidine plus water were $k_{SiD}/k_{SiH} = 0.868$ and $k_{SiT}/k_{SiH} = 0.796$ at 25°C and $k_{SiT}/k_{SiH} = 0.776$ at 0°C. With potassium hydroxide in ethanol and water at 25°C, $k_{SiT}/k_{SiH} = 0.714$ for the triphenylsilane, and for tri-n-propylsilane, 0.671. The uncertainties in these numbers are a few units in the last figure.

Using triphenylsilane and a mixture of toluene, piperidine and water as reaction medium, Brynko, Dunn, Gilman, and Hammond (22) obtained in a competitive reaction $k_{SiD}/k_{SiH} = 0.68$ and, by comparing independent runs with the isotopic substances, 0.71, both values at 25°C.

The values obtained by both groups of workers agree best with those calculated from model (II). The temperature dependence obtainable from two of the values of Kaplan and Wilzbach agree with the assumption that the rate-constant ratio could be approximated by a simple exponential, the exponent of which is inversely proportional to the absolute temperature. Thus

$$\frac{298 \log 0.796}{273 \log 0.776} = 0.98$$

a value not far from unity.

The accuracy in the values of Kaplan and Wilzbach and the fact that they have measured the deuterium and tritium isotope effects under identical conditions allow a test of Eq. (2–13); cf. pp. 23 ff. It is evident, however, that the relation cannot be expected to hold if the actual transition state is fairly similar to model (II). The reason is that the transferred hydrogen becomes attached not to a heavy atom but to a protium atom. The result of this is that the frequencies in the transition state have the ratio

$$\left(\frac{1}{1} + \frac{1}{1}\right)^{1/2} : \left(\frac{1}{2} + \frac{1}{1}\right)^{1/2} : \left(\frac{1}{3} + \frac{1}{1}\right)^{1/2} = 1 : 0.866 : 0.817$$

instead of

$$\frac{1}{\sqrt{1}} : \frac{1}{\sqrt{2}} : \frac{1}{\sqrt{3}} = 1 : 0.707 : 0.577$$

as required for the relation to hold. The silicon-hydrogen frequencies are in the ratio $1 : 0.725 : 0.600$ and thus approximately follow the idealized picture, as they should. In spite of these facts it might be of interest to carry the calculation through.

Since the symmetry numbers are cancelled, we obtain with the experimental values given above

$$\frac{\log (k_{SiT}/k_{SiH})}{\log (k_{SiD}/k_{SiH})} = \frac{\log 0.796}{\log 0.868} = 1.61$$

instead of the theoretical 1.44. Our calculated isotope effects for model (I) give nearly the expected relation:

$$\frac{-(-1282 + 2135)}{-(-1547 + 2135)} = 1.45$$

while we obtain with the calculations for model (II)

$$\frac{-(3598 - 4405 - 1282 + 2135)}{-(3817 - 4405 - 1547 + 2135)} = \infty$$

Kaplan and Wilzbach (58) also investigated the isotope effect for the hydrogen abstraction from the solvent in the same reaction. Since the stretching frequency of an OH bond is considerably higher than that of an SiH bond, the zero-point energy lost in passing into the transition state is larger for the former bond, and stronger isotope effects will arise with isotopic alcohol than with isotopic silane. A calculation according to model (I) of the effect at 25°C between tritiated and ordinary ethanol in the reaction with ordinary triphenylsilane gives, in the same manner as above and with the frequencies given there,

$$\frac{k_{ROT}}{k_{ROH}} = \exp\left[-\frac{hc}{2k \times 298}(-1988 + 3259) \right] = 0.046$$

If the isotopic ethanol molecules react with deuterated silane, nothing is changed according to this model, because the contribution from the SiD bond will be the same to both reaction rates to be compared; that is, the data of this bond are cancelled in k_{ROT}/k_{ROH}. This is not the case with model (II), because this transition state contains different isotopic molecules of hydrogen. With tritiated and ordinary ethanol reacting with deuterated silane at 25°C, the following is obtained:

$$\frac{k_{ROT}}{k_{ROH}} = \exp\left[-\frac{hc}{2k \times 298}(2846 - 3817 - 1988 + 3259) \right] =$$
$$= 0.48$$

With ordinary silane the value is $k_{ROT}/k_{ROH} = 0.33$. The observed values were 0.138 and 0.144, respectively, with deuterated and ordinary triphenylsilane. With ordinary tri-n-propylsilane, $k_{ROT}/k_{ROH} = 0.251$. The uncertainty in these values is some unit in the last figure. Since hydrogen is easily exchangeable between all positions at oxygen, the tritium content of the ethanol has simply been assumed to be equal to the average content of the ethanol, the water, and the hydroxide ions of the system. This might introduce some extra uncertainty.

In spite of the fact that the experimental isotope effect is considerably stronger than the one calculated according to model (II), the results seem to support that model. Also, the isotope effects for the isotopic silanes were found to be somewhat stronger than predicted by model (II), but of the same order of strength. Both discrepancies could be decreased if lower frequencies were assigned to the hydrogen molecule in the transition state, i.e., if the hydrogen-hydrogen bond were not considered fully developed. Such an assumption would also tend to bring the two k_{ROT}/k_{ROH} values for deuterated and ordinary silane closer together as required by the experimental figures, because in the limit represented by model (I) the difference disappears.

Although the principal discussion of complex kinetics is postponed until Chapter 6, it should be pointed out that the strong isotope effects obtained competitively with isotopic ethanol per se constitute no proof that the ethanol takes part in the reaction step which is rate-determining for the total reaction. It shows only that the slowest step with respect to ethanol involves a transition state in which a considerable part of the zero-point energy of the oxygen-hydrogen bond has been lost. This step, however, might well be subsequent to, and faster than, the one which is rate-determining for the over-all reaction. The proof, however, comes partly from the fact that an almost complete deuteration of the solvent depresses the over-all rate (58). Such an extensive deuteration of the solvent changes the medium and also the hydroxide ion attacking the silicon and is hard to treat quantitatively.

5

Secondary
Hydrogen Isotope Effects

One of the main principles of organic chemistry, particularly of the chemistry of saturated compounds, is the relative independence of different parts of a molecule in regard to chemical properties. This is a consequence of the properties of the electronic structure. To a first approximation all σ-bonds of a molecule could be treated as localized to the pair of atoms concerned. The presence of π-bonds in proper relative positions, on the other hand, gives rise to delocalization, thus opening a path for communication between different parts of the molecule. If we proceed a little beyond the very first approximation, we find also that systems of pure σ-bonds relay manifestations of different effects from one part of a molecule to another. This transmission is generally interpreted in terms of inductive and hyperconjugative effects.

The experimental results hitherto dealt with in this book have been concerned exclusively with primary isotope effects, i.e., effects caused by isotopic atoms directly involved in the reaction. The bonds opened or closed in the reaction have ended at such atoms. In view of these statements, however, it is not astonishing to find that atoms not directly occupied in bond opening or bond making could influence the reaction rate by their isotopic mass.

If we consider the general expressions for the isotope effect, e.g., Eq. (2–5) or (2–16), it is obvious that nothing directly requires the isotopic atom to be located at a reacting bond in order for it to cause an isotope effect. It is only by virtue of its influence on the molecular masses, moments of inertia, and/or vibrational frequencies that the isotopic mass of an atom might influence the reaction rate. In practical applications of the equations, however, it becomes clear that most of this influence is cancelled unless the isotopic atom is not involved in the reaction. This is easily seen if the crude model with valence-bond oscillators is considered. The frequencies are changed upon isotopic substitution only for those bond oscillators which have the isotopic atom at one end. These changes will generally be cancelled between the reactant and the transition state if the frequencies are not too different in these two entities. What generally is not cancelled, on the other hand, is the stretching frequency of a bond going to be ruptured. This movement, which is no longer periodic in most transition states, is excluded from the partition function of the latter, and the corresponding periodic motion in the reactant is left uncancelled. In order for an isotopic atom to cause a primary isotope effect, it is obviously necessary that it be directly connected with the reacting bond. On the other hand, if the vibrational patterns of the reactant and the transition state do not match perfectly in some other part of the molecule (this, of course, will always be found if the claim for accuracy is made rigorous enough), isotopic atoms in those favorable parts get a chance to cause a secondary isotope effect. In the first place the isotopes of hydrogen could be expected to give such effects of measurable magnitude owing to their large relative mass difference.

Recent discussions concerning the relation between hydrogen isotope effects and hyperconjugation, which are of considerable interest in the present connection, have been published by Lewis (61) and Shiner (91).

SECONDARY HYDROGEN ISOTOPE EFFECTS IN ALIPHATIC SOLVOLYTIC REACTIONS

A valuable survey of what is known experimentally of secondary isotope effects in aliphatic solvolytic reactions, together with a theoretical treatment, has recently been given by Streitwieser, Jagow, Fahey, and Suzuki (93), who simultaneously present some results of their own. Carbonium ion formation from aliphatic derivatives containing hydrogen in α- or β-position generally exhibits quite measurable isotope effects, the heavy compound having the lowest reaction rate as usual.

For the same reasons as in primary hydrogen isotope effects the zero-point energy factor (Eq. 2–9) is probably decisive. The three vibrational modes of each isotopic carbon-hydrogen oscillator should be included, and it should be remembered that none of these vibrational degrees of freedom vanishes in the transition state. Owing to the small mass of hydrogen, we have for all three frequencies approximately

$$\frac{\nu_1}{\nu_2} \approx \left(\frac{m_2}{m_1}\right)^{1/2}$$

where m denotes hydrogen mass. Since the carbon mass is not very large compared to that of hydrogen, Streitwieser and co-workers propose to use an empirical ratio, somewhat closer to unity than $(m_2/m_1)^{1/2}$. For the couple deuterium/protium they write

$$\frac{\nu_D}{\nu_H} = \frac{1}{1.35}$$

instead of the usual $1/\sqrt{2}$. The zero-point-energy factor now becomes

$$\frac{k_D}{k_H} = \exp\left[-\frac{1}{2}\left(\frac{1}{1.35} - 1\right)\sum_i \left(u_{i(H)}^{\ddagger} - u_{i(H)}\right)\right] =$$

$$= \exp\left[\frac{0.130hc}{kT}\sum_i \left(\bar{\nu}_{i(H)}^{\ddagger} - \bar{\nu}_{i(H)}\right)\right]$$

where the sum is to be taken over all vibrations of isotopic carbon-hydrogen oscillators (the same number in the reactant and the transition state), and $\bar{\nu}$ denotes wave number. All other frequencies are assumed to be cancelled. The symmetry numbers have been omitted. A normal isotope effect with the heavy compound reacting more slowly requires the frequencies of the transition state to be lower than those of the reactant on the average.

Streitwieser and co-workers determined the rate of acetolysis of the toluene-p-sulfonates (tosylates) of cyclopentanol, cyclopentanol-1-d, cis- and $trans$-cyclopentanol-2-d and cyclopentanol-2,2,5,5-d_4 in separate runs at 50°C. The medium was dry acetic acid containing sodium acetate in slight excess over the tosylate. The reaction is known to be of the S_N1 type, the rate-determining step being the formation of the carbonium ion. The k_D/k_H ratio was equal to 0.87 when the deuterium was situated in the α-position. With one deuterium atom in $trans$-β-position, $k_D/k_H = 0.86$ and, with one in cis-β-position, 0.82. When all four β-positions were occupied by deuterons, $k_D/k_H = 0.49$. The figures for β-substitution show that each deuterium atom introduced changes the rate by the approximate factor 0.84, fairly irrespective of whether the position is cis or trans relative to the tosylate group. This figure is also fairly representative of the general strength of these secondary isotope effects. According to the expression above, such effects require the sum of the vibrational frequency differences to be about -240 cm^{-1} for the α-position and -300 cm^{-1} for the β-position per deuterium atom introduced. The question is now whether these frequency depressions in the transition state are compatible with reasonable assumptions about the structure of the transition state. We shall see that they are indeed compatible.

α-Deuterium Effect. The cyclopentanol derivatives were chosen in order to minimize effects caused by conformational differences. The cyclopentane ring is not too puckered, and the cyclopentyl cation is probably fairly flat, with the α-hydrogen approximately in the ring plane, the hybridization of the positively charged carbon atom being sp^2. In the reactant, the carbon-hydrogen frequencies are certainly those generally occurring in similar positions, 2890 cm^{-1} for stretching and 1340 cm^{-1} for each of two bending modes (3a). In order to obtain a model for the carbon-hydrogen oscillator at the positively charged carbon of the transition state, Streitwieser et al. compared it to the one in aldehydes, where the hybridization at carbon is the same and there is probably at least some positive charge on the carbon. From different sources they obtained the average frequencies 2800 cm^{-1} for stretching, 1350 cm^{-1} for bending in plane, and 800 cm^{-1} for bending out of plane. The conclusion is obviously that it is mainly the decrease in frequency of one of the bending modes that causes the α-isotope effect. A fraction of the total frequency depression in the formation of the carbonium ion is enough to explain the effect, and it seems probable that the leaving tosylate ion still has some influence in the transition state.

In this connection it is interesting to note that a corresponding $S_N 2$ reaction has been found to give no α-deuterium isotope effect. Shiner (88) found that 2-bromopropane-2-d reacted with sodium ethoxide with the same velocity (within 1 per cent) as did the protium compound. In this case the transition state certainly contains the ethoxide ion and the bromide ion half-bound on each side of the approximately flat [CH(CH$_3$)$_2$]$^+$ residue. Their presence is likely to increase the out-of-plane vibration frequency, thus compensating the decrease that takes place in a freer carbonium ion. These views presented by Streitwieser et al. have received extremely good support from Johnson and Lewis (53), who have compiled results of their own and others and have shown that increasing necessity for nucleophilic attack from the rear in a reaction is accompanied by a weakening of the α-deuterium isotope effect. Increasing necessity for nucleophilic attack is presumably also accompanied by decreasing distance between the α-hydrogen and the nucleophile.

β-Deuterium Effect. For the β-deuterium effect, a possible solvation of these hydrogens in the transition state, which certainly exists in elimination reactions, has been taken into consideration as a source of the isotope effect. If the transition state in the acetolysis of cyclopentyl tosylate is actually fairly similar to a flat cyclopentyl cation, there is practically only one conformation. The presence of a tosylate group on one side of the ring plane would interfere differently with solvating species operating on the *cis*- and on the *trans*-hydrogen. The facts that the *cis*- and the *trans*-β-deuterium give about the same isotope effect, and that the *trans* gives the weaker one, make such hypotheses less probable.

Streitwieser *et al.* prefer hyperconjugation as the explanation of the β-deuterium effect. For example, in a flat cyclopentyl cation there ought to be considerable hyperconjugation between the empty *p*-orbital at carbon, which has its axis perpendicular to the ring plane, and the orbitals of neighboring carbon-hydrogen bonds. The orbitals are not parallel, but they are still in a favorable mutual position. When the axes of two such orbitals are at right angles, hyperconjugation is inhibited. In the reactions with the cyclopentyl tosylates the geometrical conditions for hyperconjugation in the transition state are thus fairly well fulfilled. There exists experimental evidence for a considerable weakening of the effect with molecules which cannot acquire the proper conformation for steric reasons. Thus Shiner (90) has found a considerably weaker effect in the solvolysis of 2,4,4-trimethyl-2-chloropentane-3,3-d_2,

$$
\begin{array}{ccccc}
& CH_3 & & CH_3 & \\
& | & & | & \\
CH_3 & —C— & CD_2— & C— & CH_3 \\
& | & & | & \\
& Cl & & CH_3 &
\end{array}
$$

than in that of the 1,1,1-d_3 derivative,

$$
\begin{array}{ccccc}
& CH_3 & & CH_3 & \\
& | & & | & \\
CD_3 & —C— & CH_2— & C— & CH_3 \\
& | & & | & \\
& Cl & & CH_3 &
\end{array}
$$

The isotope effects per deuteron were $k_D/k_H = 0.96$ and 0.89, respectively, at $25°C$ in aqueous ethanol.   The latter figure is comparable to those obtained (89) for the molecules

$$\underset{\displaystyle \text{Cl}}{\overset{\displaystyle \text{CH}_3}{\text{CH}_3\text{—C—CD}_2\text{—CH}_3}} \quad \text{and} \quad \underset{\displaystyle \text{Cl}}{\overset{\displaystyle \text{CD}_3}{\text{CD}_3\text{—C—CH}_2\text{—CH}_3}}$$

under the same conditions: $k_D/k_H = 0.84$ and 0.91, respectively, per deuteron.

Hyperconjugation involves overlapping between the β-carbon-hydrogen bonding orbital and the empty p-orbital at the α-carbon.   The electron cloud initially concentrated around the β-carbon-hydrogen axis is extended toward the α-carbon orbital. This tendency of the electron cloud to become extended certainly weakens the force opposing a movement of the hydrogen in the same direction, and the frequency of the corresponding vibrational mode decreases.   In this way Streitwieser *et al.* ascribe the β-deuterium effect to hyperconjugative decrease of the frequency of a bending vibrational mode of these hydrogens.

Applications. From what has been said above about secondary hydrogen isotope effects, it is evident that they are quite measurable and should be useful as a tool in the study of transition states.   For instance, the fact that the α-hydrogen effect is sensitive to the presence of an incoming nucleophilic entity means that measurements of such effects should be excellent diagnostic tools for following the transition from S_N1 to S_N2 reactions.   (Cf. also the discussion of the carbon isotope effect in similar reactions, pp. 143 ff.)

SECONDARY ISOTOPE EFFECTS CAUSED BY ALIPHATIC–AROMATIC HYPERCONJUGATION

It is generally assumed that hyperconjugation exists between the π-electron system of a benzene ring and the α-carbon-hydrogen bonds of an aliphatic substituent. In view of the facts discussed in the previous section, it could then also be expected that substitution of isotopic hydrogen in these positions should be able to cause secondary isotope effects in those reactions for which there is a difference in hyperconjugative power between the reactant and the transition state. Conversely, measurements of such effects ought to tell something about the extent of hyperconjugation. If in the pertinent reactions the center of reaction is situated in para position relative to the substituent carrying the isotopic hydrogen, disturbances caused by the mere proximity of this center are less likely to play an important role.

Solvolysis of p-Alkylphenylcarbinyl Chlorides. Lewis and Coppinger (62) have studied the first-order rates of acetolysis of differently deuterated methyl-p-tolylcarbinyl chlorides,

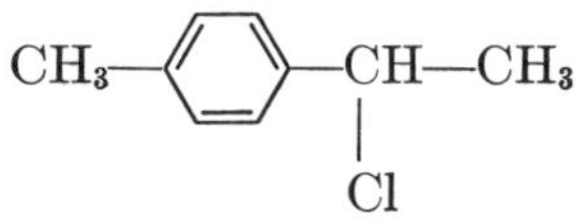

$$CH_3-\phenyl-CH-CH_3$$
$$|$$
$$Cl$$

and have found that deuterium substitution in the methyl group to the left causes a rate change by a factor $k_D/k_H = 0.96$ per deuterium atom at 50°C, and 0.97 at 65°C, the deviations from unity being quite beyond the experimental uncertainty. The corresponding figures for deuterium substitution in the methyl group to the right were 0.90 and 0.94. It is thus evident that the remote methyl hydrogen is able to give rise to a measurable secondary isotope effect. Probably the effect is relayed by the hyperconjugative mechanism, which for the final cation might be pictured

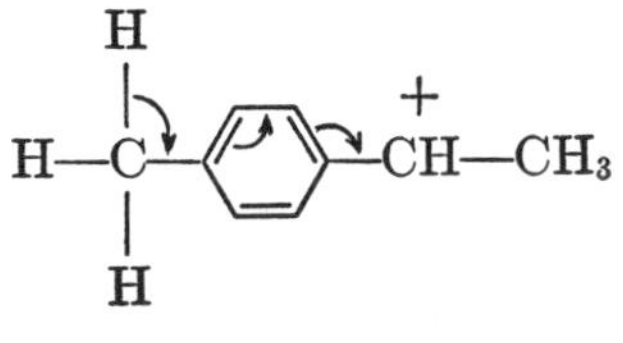

$$H-\underset{\underset{H}{\overset{\overset{H}{|}}{|}}{C}-\phenyl-\overset{+}{C}H-CH_3$$

Similar results have been obtained by Shiner and Verbanic (92) in an investigation of the solvolysis in aqueous ethanol and aqueous acetone of benzhydryl chlorides

carrying a *p*-alkyl group R with isotopic hydrogen in the α-position. The effects were rather weak and showed solvent dependence. At 0°C the total effect of changing R from CH_3 to CD_3 was $k_D/k_H = 0.945$ in 80% acetone, and less in other solvents. Deuteration in the α-position of the groups R equal to CH_3CH_2, $(CH_3)_2CHCH_2$, and $(CH_3)_2CH$ gave smaller effects, decreasing in the same order.

Electrophilic Aromatic Substitution. The role of hyperconjugation of a methyl group, for example, with the aromatic nucleus in electrophilic aromatic substitution is still far from settled. This fact makes investigations of corresponding secondary isotope effects so much more interesting and important. Swain, Knee, and Kresge (94) report, for the nitronium ion nitration of toluene and toluene-α-t, $k_T/k_H = 0.997 \pm 0.003$; for mercuration of toluene-α,α,α-d_3, $k_D/k_H = 1.00 \pm 0.03$ per deuterium atom; and, for the Br_2 bromination of toluene-α-t, $k_T/k_H = 0.956 \pm 0.008$; all reactions being carried out at 25°C. Thus the real effects, which possibly exist, are small. It might be significant that bromination with molecular bromine, which is known to be very sensitive to directing effects, gives the single definitively established effect.

In a very recent personal communication, Professor W. M. Lauer reports a value of $k_D/k_H = 0.95$ for the over-all effect of introducing three methyl deuterons in toluene on the rate of deuteration of the molecule in trifluoroacetic acid at 70°C. This must certainly also be considered a surprisingly weak effect.

ISOTOPE EFFECTS OF OTHER ORIGIN

In general, one must be wary of drawing far-reaching conclusions from very small isotope effects the origin of which might be doubtful.

Anharmonicity. Halevi (43) has drawn attention to the fact that vibrational anharmonicity causes the carbon-hydrogen (average) distance to be somewhat different for the different hydrogen isotopes (even in the vibrational ground state). Thus an increase in the atomic mass of the hydrogen is accompanied by a decrease in the bond length, and consequently by an increase in electron density at the carbon atom. This is equivalent to an increase in the inductive electron release from the carbon-hydrogen bond to other parts of the molecule. There are experimental facts which could be interpreted in this manner, although the question of the magnitude of the effect is still far from settled; cf. papers by Streitwieser *et al.* (93), Lewis (61), and Shiner (91).

Halevi claims that the inductive effect of a C—D bond relative to that of a C—H bond is large enough to obscure the relation between hyperconjugation and isotope effects. Thus, for instance, the vanishing secondary isotope effect in aromatic substitution (previous section) could be considered a result of a cancellation between a retarding hyperconjugative effect of the introduction of deuterium and an accelerating inductive one. These interesting problems need and deserve further studies.

"False" Isotope Effects. It might be instructive to look for the condition when two isotopic substances will react with exactly the same rate. From Bigeleisen's complete formula, Eq. (2–16), it is evident that only a complete detailed matching of the frequencies, either of type $\nu_{i(1)}^{\ddagger} = \nu_{i(2)}^{\ddagger}$, $\nu_{i(1)} = \nu_{i(2)}$ or of type $\nu_{i(1)}^{\ddagger} = \nu_{i(1)}$, $\nu_{i(2)}^{\ddagger} = \nu_{i(2)}$, can warrant a non-fortuitous absence of isotope effect. Equations (2–14) and (2–15) tell that under such circumstances the following relationship must hold:

$$\left(\frac{M_1}{M_2}\right)^{3/2} \left(\frac{A_1B_1C_1}{A_2B_2C_2}\right)^{1/2} = \left(\frac{M_1^{\ddagger}}{M_2^{\ddagger}}\right)^{3/2} \left(\frac{A_1^{\ddagger}B_1^{\ddagger}C_1^{\ddagger}}{A_2^{\ddagger}B_2^{\ddagger}C_2^{\ddagger}}\right)^{1/2}$$

This relation will, of course, seldom be exactly satisfied. For instance, in a unimolecular decomposition, $M_1 = M_1^{\ddagger}$ and $M_2 = M_2^{\ddagger}$, and it is required that the two moment-of-inertia ratios be equal. Such a relation holds for a diatomic molecule, but certainly not for molecules in general. Suppose, for example, that the molecule initially has the isotopic atom at its center of gravity. The left-hand ratio is then unity. If passing into the transition state involves non-symmetric deformation, the isotopic atom cannot be situated in the center of gravity of the transition state, and the right-hand ratio must differ from unity.

It seems hard to put a limit to the strength, but, even if the isotopic atom is well protected against any intramolecular influence from the center of reaction, complete absence of an isotope effect should be the exception rather than the rule. This is a consequence of mechanical principles rather than one of chemical relay mechanisms.

6

Hydrogen Isotope Effects and Reaction Mechanisms

The previous discussions of hydrogen isotope effects dealt mainly with reactions the mechanisms of which were not too complicated and were more or less known from independent evidence. The primary purpose of these discussions was to show how strong isotope effects can generally be expected for single reaction steps. Obviously, purely theoretical predictions are unable to give more detailed information about the strength of the effects. On the other hand, one must admit that there is no reason to suspect the theory to be wrong, for experimental isotope effects will seldom exceed such limits as could be assigned to the strength by means of the theory and reasonable assumptions. The causes of the shortcomings of our present theory are to be sought in inadequate input data. Nothing will prevent us from using isotope effects in an empirical, semiquantitative, or qualitative fashion as a diagnostic tool in order to gain evidence of the course of chemical reactions of different degrees of complexity. Experience has already shown the power of kinetic isotope effects as a complement to more classical reaction kinetics in those cases where the latter is unable to give an unequivocal answer.

Although several of the experiments already discussed in this book could also have served as examples of investigations of reaction mechanisms, and the boundary between the present chapter and Chapter 4 is vague, the present one has been written to provide a few illustrations of how hydrogen isotope effects have been used as an auxiliary means in research work on reaction mechanisms.

METALATION OF AROMATIC COMPOUNDS

There have been two main hypotheses for the mechanism of the metalation of aromatic compounds. Only the basic ideas of these hypotheses, not the arguments in favor of either of them, will be reviewed here.

According to one of them, the crucial step consists of an electrophilic attack of the metal cation of the organometallic reagent on the aromatic carbon atom, from which the hydrogen is subsequently, and probably easily, expelled. Such a mechanism is equivalent to that in ordinary electrophilic aromatic substitution, which will be discussed later in the chapter. Since it has been shown that most electrophilic aromatic substitutions proceed without measurable hydrogen isotope effect, it would seem natural to expect a similar behavior of the metalation.

The alternative mechanism involves a primary nucleophilic attack on the hydrogen by the carbanion of the organometallic reagent. It is probably also of great importance for the metal atom to have an opportunity to coordinate to a neighboring hetero atom if possible. The abstraction of the proton is likely to be rate-determining and cause an appreciable isotope effect (cf. Chapter 4).

Bryce-Smith, Gold, and Satchell (21) have measured the deuterium isotope effect in the metalation of deuterobenzene and toluene-α-d with ethylpotassium. The deuterium content of the product (after conversion to the corresponding carboxylic acid) was compared to that of the reactant. Toluene, which is substituted in the side chain, gave the value $k_D/k_H = 0.22 \pm 0.05$ at 20°C. This could, of course, be expected as the saturated side chain offers no possibility of an addition step.

Benzene-d was metalated at 75°C, giving $k_D/k_H = 0.50 \pm 0.10$. Thus metalation in an aromatic nucleus also proceeds with a measurable, although not strong, isotope effect. Bryce-Smith, Gold, and Satchell point out that the tunnel effect might also be operative to some extent.

A similar investigation has been undertaken by Gronowitz and Halvarson (41), who metalated thiophene, containing tracer amounts of thiophene-2-t, with n-butyllithium. The metalation reaction is directed exclusively toward the positions adjacent to the sulfur atom, probably owing to coordination of the reagent to the latter. The result was $k_T/k_H \leq 0.17$ at about room temperature.

The absence of any appreciable isotope effect in the metalation of aromatic nuclei would have proved conclusively that the reaction is analogous to nitration and most other electrophilic aromatic substitutions. The effect which has been established, however, is a strong support of the primary nucleophilic hydrogen abstraction, although, it must be admitted, a less likely possibility would be the electrophilic substitution mechanism with the second step, the final elimination of the proton, rate-determining (cf. pp. 107 ff.). However, the isotope effect taken together with other experimental evidence, for instance the directional properties of the reaction, offers sufficient evidence for the primary proton abstraction.

OXIDATION OF ALCOHOLS

Oxidation with Chromic Acid. The oxidation of alcohols includes a great variety of reactions, and the mechanisms are frequently very complicated. The method of using isotope effects to find the rate-determining step was used very early by Westheimer and his co-workers, who investigated particularly the chromic acid oxidation of isopropyl alcohol. Owing to the numerous valence states of chromium, the mechanism is very complex and involves more than one single oxidation step. It is not possible to discuss the over-all reaction scheme here, but reference should be made to the classical paper of Westheimer and Nicolaides (99), who found that 2-propanol-2-d was reduced at a rate of about one-sixth that of ordinary isopropyl alcohol at 40°C. This showed that the secondary hydrogen is removed in the rate-determining, first step. Since there was reason to assume that monoisopropyl chromate is an intermediate, Leo and Westheimer (60) later investigated chromic acid esters. They showed that the rate of the internal oxidation-reduction of diisopropyl chromate in benzene is depressed to about a fifth by deuterium substitution at the secondary carbon atom. Subsequent work by Kaplan (56), who measured the relative rates of oxidation of 2-propanol and 2-propanol-2-t competitively, makes it seem probable that the rupture of the tertiary carbon-hydrogen bond is rate-determining also for that faster part of the over-all reaction which involves oxidation by Cr(IV), even if the possibility that a minor fraction follows another path could not be completely excluded.

Oxidation with Bromine. The somewhat simpler reaction between ethanol and bromine, which has been investigated by isotopic means by Kaplan, will be discussed in greater detail.

From earlier, classically kinetic work, not to be reviewed here, it was known that in dilute aqueous solution the over-all reaction consists of two irreversible consecutive steps, the first of which is the slower:

$$CH_3CH_2OH + Br_2 \xrightarrow[(slow)]{} CH_3CHO + 2HBr$$

$$CH_3CHO + Br_2 + H_2O \xrightarrow[(fast)]{} CH_3COOH + 2HBr$$

The second step is about 200 times faster than the first one. The over-all rate, which is nearly equal to the rate of the first step (except for an initial period), had been found to be first-order in ethanol and first-order in bromine. From this it is evident that one molecule of ethanol and one molecule of bromine have to come together to produce acetaldehyde. The problem to be solved by isotopic means is the more detailed mechanism. The most attractive picture of the first step seemed to be

$$CH_3CH_2OH + Br_2 \xrightarrow[(slow)]{} CH_3CH_2OBr + HBr$$

$$CH_3CH_2OBr \xrightarrow[(fast)]{} CH_3CHO + HBr$$

because alkyl hypochlorites are known to behave as in the second of these reactions.

If the rate-controlling attack is on the hydroxyl hydrogen, an ethanol molecule with heavy hydrogen in the methylene group should react at very nearly the same rate as ordinary ethanol. Kaplan (55) made experiments with ethanol containing tracer amounts of ethanol-1-t and followed the specific activity of the unreacted alcohol. In this case we have the reaction scheme

$$CH_3CH_2OH + Br_2 \xrightarrow{k_1} CH_3CHO + 2HBr$$

$$CH_3CHTOH + Br_2 \begin{cases} \xrightarrow{k_2'} CH_3CHO + TBr + HBr \\ \\ \xrightarrow{k_2''} CH_3CTO + 2HBr \end{cases}$$

From Eq. (3–4) it was possible to calculate the value $(k_2' + k_2'')/k_1$ from the experimental data. It was found to be about 0.575 at 37.5°C. For approximate purposes it could be assumed that $2k_2'' \approx k_1$; hence $2k_2'/k_1 = 0.15$. Such a value is not compatible with an attack on the hydroxyl hydrogen but rather a proof that the attack is on a methylene hydrogen.

If the hypobromite is not a precursor of the aldehyde, it is still possible to think that the rate-determining step is the formation of hypobromite, which then in a fast reaction attacks another ethanol molecule in methylene position. This means that the rate-determining step, although it has to do with the ethanol, should be a mere production of a reagent, and it could have no discriminatory influence on ethanols isotopic in the methylene position. The discrimination between the isotopic molecules comes in the second step, which, although not over-all rate-determining, involves the actual competition between the isotopic molecules. Isotope effect techniques offer a possibility to check this in another way. If separate runs are arranged with ordinary ethanol and ethanol-1,1-d_2, the over-all rates are measured, and they should be practically equal provided the hypobromite production is rate-controlling.

Kaplan (57) also arranged experiments of the latter kind. These non-competitive experiments gave a value of $k_D/k_H = 0.23$ at 24.8°C. It is obvious that the second hypothesis involving ethyl hypobromite has to be discarded also, as the deuterium isotope effect is quite comparable to the tritium one. In order to see this we use Eq. (2–13) and find that k_T/k_H ought to be $(0.23)^{1.44} = 0.12_0$ at the same temperature, 24.8°C. Since we assume a pure zero-point-energy effect, the logarithm of k_T/k_H could also be assumed to be inversely proportional to the absolute temperature. Such a temperature correction gives $k_T/k_H = 0.13$ at 37.5°C, which is equal to the above value within the limits of accuracy.

Also, the possibility of the formation of ethyl hypobromite in a rapid reversible reaction preceding the rate-determining step must be discarded, since the position of the equilibrium

$$CH_3CH_2OH + Br_2 \rightleftarrows CH_3CH_2OBr + HBr$$

depends on the growing concentration of hydrobromic acid, and such dependence of the reaction rate is contrary to fact under the conditions of the experiment.

The most natural explanation of the experimental facts seems to be a mechanism consisting of a direct and rate-controlling attack on the methylene group by the bromine molecule, for instance,

$$CH_3CH_2OH + Br_2 \longrightarrow [CH_3CHOH]^+ + HBr + Br^-$$

A possible rapid pre-equilibrium,

$$CH_3CH_2OH + Br_2 \rightleftarrows CH_3CH_2OH \cdot Br_2$$

is evidently kinetically insignificant. Such a mechanism, with or without a pre-equilibrium, would certainly give the observed isotope effects.

Competitive versus Independent Measurements. In the investigation discussed above a very important kinetic principle for isotopic reactions has been used. Obviously the two measurements of the isotope effect, one competitive and the other consisting of independent runs with macroscopic amounts of the isotopic molecules, would not have given comparable results if the production of reagent hypobromite had been rate-determining. The general conclusion to be drawn is that competitive and non-competitive measurements do not necessarily give the same result in reactions composed of several steps.

Independent runs with isotopically pure species will, of course, always tell the rate of the over-all rate-determining step and hence its isotope effect. Competitive reactions, on the other hand, are able to give information about the isotope effect in the reaction step which is decisive for the choice between different isotopic species. The latter step need not be rate-determining for the over-all reaction, but that role might, for instance, be played by a reaction generating a necessary reagent. As in the present example, these principles are very helpful in the elucidation of reaction mechanisms, but they certainly also call for great care in the discussion of a measured isotope effect when the course of the reaction is not known in detail. Particularly if consecutive reactions have comparable rates, the resulting isotope effect might acquire values characteristic of none of the elementary reactions.

An illustration of a reaction that gives different isotope effects according to the two methods is the oxidation of 2-propanol with chromic acid, mentioned above. Independent runs tell only the rate of the rate-controlling first step, which involves oxidation by hexavalent chromium. As soon as the reduction of chromium from its hexavalent state has been initiated, it is then rapidly completed down to the trivalent state, and it is insignificant for the measured rate whether these subsequent reactions occur with heavy or light alcohol molecules. The observed isotope effect is representative exclusively of the first step. On the other hand, in a competitive measurement there is a choice between the two isotopic molecular species in the subsequent steps as well as in the first one. The measured isotope effect will be a weighted mean of the partial ones. In the case with chromic acid oxidation, the partial isotope effects happened to be different (56), and the mean isotope effect differed appreciably from that of the first step, which probably consumes only a third of the reacted propanol.

Oxidation with Diazonium Ions. The results obtained in the two oxidation reactions described are closely related to those of Melander (65) concerning the reaction between nitrobenzene-diazonium ions and tritiated samples of methanol and ethanol. It was shown that it is one of the hydrogens at carbon atom 1 which is incorporated into the produced nitrobenzene and that k_T/k_H was approximately $\frac{1}{7}$ at about 350°K, as computed from measurements on the initial alcohol and the aromatic product. In the light of the later investigation of this type of reaction carried through by DeTar and Turetzky (30), it seems certain that the pertinent reaction step is of radical character:

$$C_6H_5\cdot + CH_3OH \longrightarrow C_6H_6 + \cdot CH_2OH$$

or, in the case of ethanol:

$$C_6H_5\cdot + CH_3CH_2OH \longrightarrow C_6H_6 + CH_3\overset{\cdot}{C}HOH$$

i.e., a hydrogen atom abstraction from carbon atom 1 by phenyl radicals.

DECOMPOSITION OF p-NITROPHENETHYL-TRIMETHYLAMMONIUM ION

The generally accepted mechanism for the formation of olefins from tetraalkylammonium ions, like the p-nitrophen-ethyltrimethylammonium ion, is a concerted one, initiated by the attack of a nucleophilic species, B:, on the β-hydrogen:

There are, however, related reactions which proceed according to

$$HCR_2\text{—}CR_2\text{—}X \xrightarrow[\text{(slow)}]{} HCR_2\text{—}\overset{+}{C}R_2 + X^{\checkmark}$$

$$HCR_2\text{—}\overset{+}{C}R_2 \xrightarrow[\text{(fast)}]{} H^+ + R_2C\text{=}CR_2$$

where X could be, for instance, the group $\overset{+}{S}R_2$. In order to confirm the first mechanism in the case of p-nitrophenethyltrimethylammonium ion, Hodnett and Flynn (51) decomposed a sample of the iodide containing tracer amounts of tritium at the β-position in aqueous solution at 100°C. The reaction scheme is

$$B + O_2N \cdot C_6H_4 \cdot CH_2CH_2\overset{+}{N}(CH_3)_3$$
$$\xrightarrow{k_1} O_2N \cdot C_6H_4 \cdot CH\text{=}CH_2 + BH^+ + N(CH_3)_3$$

$$B + O_2N \cdot C_6H_4 \cdot CHTCH_2\overset{+}{N}(CH_3)_3$$

$$\begin{cases} \xrightarrow{k_2'} O_2N \cdot C_6H_4 \cdot CH\text{=}CH_2 + BT^+ + N(CH_3)_3 \\ \xrightarrow{k_2''} O_2N \cdot C_6H_4 \cdot CT\text{=}CH_2 + BH^+ + N(CH_3)_3 \end{cases}$$

The change in the specific activity of the reactant during the reaction gives $(k_2' + k_2'')/k_1$ according to Eq. (3–4).

At small conversions, i.e., within the "linear region" discussed on pp. 57 ff., the amount of conversion of each isotopic molecular species is proportional to the specific rate (Eq. 3–17), in our case

$$\frac{x_1}{x_2} = \frac{k_1}{k_2' + k_2''}$$

Since the fraction $k_2''/(k_2' + k_2'')$ of the nitrostyrene produced from the heavy compound contains tritium, the ratio r of the molar activities of the product and the initial reactant could be written

$$r = \frac{k_2''}{k_2' + k_2''} \times \frac{x_2}{x_1} = \frac{k_2''}{k_1}$$

[This expression could, of course, also have been obtained from Eq. (3–8), which in this case is written (observe changed indexing)

$$r = \frac{k_2''}{k_2' + k_2''} \times \frac{1}{x_1}\left[1 - (1 - x_1)^{(k_2' + k_2'')/k_1}\right]$$

Binomial expansion gives the above expression for small values of x_1.] Comparison of the molar activity of the initially produced nitrostyrene with that of the reactant thus gives the ratio k_2''/k_1, and the relative magnitude of all three specific rates can be computed from the two measurements. The result was $k_2'/k_2'' = 0.16 \pm 0.06$ and $2k_2''/k_1 = 0.87 \pm 0.04$.

From the figures obtained it is evident that the rupture of the carbon-hydrogen bond takes place in the rate-determining step of the reaction. The presence of a tritium atom in the β-position also slows down the splitting off of the neighboring protium. This secondary isotope effect is certainly related to those discussed in Chapter 5. Owing to the low frequency of the out-of-plane carbon-hydrogen vibrations of olefins (3b), it is easy to explain the effect as a result of zero-point-energy loss in passing into the transition state.

If the slow step of the reaction had been a solvolytic cleavage of the carbon-nitrogen bond, the ratio $(k_2' + k_2'')/k_1$ would have shown merely a secondary isotope effect instead of being about 0.5.

ELECTROPHILIC AROMATIC SUBSTITUTION

There are two main mechanisms which could be taken into consideration for the course of electrophilic aromatic substitution. Since aromatic molecules are known to be considerably stabilized by their particular type of π-electron resonance, it would not be unnatural to think of the substitution process as one of direct replacement. According to Ingold's notation it could be characterized as S_E2 or S_E3, depending on whether the transition state is best described as excluding or including the base that finally takes care of the proton expelled. Such a transition state could be depicted in the following way:

$$\left[\begin{array}{c} \hexagon \overset{H}{\underset{X}{<}} \end{array} \right]^+$$

where X^+ is the attacking entity and possible basic species
have been left out. The entering and leaving substituents
would probably be situated in a plane perpendicular to that
of the ring and cutting the latter along a line through the
carbon atom at the reaction center and the carbon atom in
the para position. The π-electron cloud could be imagined
to be polarized but not broken up, thus requiring no energy
to destroy the resonance. The hybridization at the carbon
atom in the reaction center would be sp^2 all the time, and H^+
would simply be pushed out of its bond by X^+. There is, of
course, no possibility of an inversion of the organic residue
similar to that occurring with aliphatic residues in S_N2 reactions. This fact makes the situation energetically less favorable than in the aliphatic case, and one might question whether
the electronic structure just described is really the most stable
possible for the arrangement of atoms concerned. It seems
almost certain that this atomic arrangement has to be passed
through, and the electronic structure realized by nature is
certainly the one of maximum stability.

As an alternative to the electronic structure described above,
one could conceive that the carbon atom at which substitution
occurs acquires sp^3 hybridization, allowing H as well as X to
be bound by ordinary σ-bonds. The π-electron structure
typical of benzene is destroyed, and the four remaining π-elec-
trons will have to distribute themselves over the five remaining
carbon p-orbitals. Owing to symmetry conditions, however,
some of the energy expended in breaking up the aromatic
structure might be regained in hyperconjugation between the
remaining π-electron system and the C—H bond (also, when
possible, the C—X bond) outside the ring plane.

If the last-mentioned alternative is realized, the atomic
arrangement under discussion is more likely to have the character of an intermediate than that of a transition state, because
there will probably be some force opposing the splitting off

of either of the ligands H and X, which are now held by full
σ-bonds. If this reaction path actually offers a more favorable
way than the other, the substitution could best be described
as a two-stage reaction:

where any basic species accepting the proton has been left out.

The rates will be determined by transition states on either
side of the intermediate, and the one corresponding to the
highest energy will determine the over-all rate. The transition
states, being defined as the complexes of maximum potential
energy that have to be passed through, will come at those
points on the reaction coordinate where the expenditure of
energy for the breaking up of the π-electron system is just
balanced by the gain from the incipient σ-bond formation and
possible hyperconjugative effects.

By kinetics in the classical sense it is not possible to decide
the question, because both mechanisms involve simply that the
aromatic molecule and the substituent have to come together.
The influence of other substituents on the reaction rate is also
unable to give information in this respect. These questions
have been discussed at some length by Melander (67). Studies
of solvent effects on reaction rate led Hughes, Ingold, and Reed
(52) to conclude that the two-stage mechanism with the first
step rate-determining was realized in the nitration with nitronium ions.

Hydrogen isotope effect studies, however, offer much more
direct evidence under favorable conditions. The one-stage
S_E2 or S_E3 mechanism should, of course, give rise to an appreciable isotope effect since the carbon-hydrogen bond is
ruptured in the concerted process. The same should apply
to the second step of the two-stage mechanism. The first
step of the latter mechanism, however, could not be expected
to give more than some secondary isotope effect, because the
carbon-hydrogen bond is certainly changed somewhat but

not ruptured. Provided the two-stage mechanism is real and the first step is rate-determining, the over-all reaction should have such a weak isotope effect. These different possibilities have been discussed by Melander (64).

Nitration. Many electrophilic aromatic substitution reactions have been found to be without measurable hydrogen isotope effect. Melander (64), using tracer amounts of tritium in competitive reactions, investigated the nitronium ion nitration of benzene, toluene, nitrobenzene, bromobenzene, p-nitrobromobenzene, and naphthalene and found that $k_T/k_H > 0.74$ in all cases; for some of them this value was considerably closer to unity. These results have been corroborated by Lauer and Noland (59), who mono- and dinitrated a mixture of monodeutero- and ordinary benzene, and by Bonner, Bowyer, and Williams (17), who nitrated nitrobenzene-d_5 and ordinary nitrobenzene in independent runs under highly acid nitrating conditions. Under such conditions, if ever, the final proton expulsion should be a slow process, and it is remarkable that the completely deuterated molecule was found to be nitrated with the same rate as the ordinary one (within 1 to 5%).

Halvarson and Melander (44) obtained $k_T/k_H = 1.0 \pm 0.1$ in the nitration of anisole with benzoyl chloride plus silver nitrate in acetonitrile. This result rules out a possible concerted mechanism,

which has been proposed for similar reactions.

Dr. Ph. C. Myhre reports in a personal communication that he has studied the nitration of 1,3,5-tri-*tert*-butylbenzene containing tracer amounts of tritium in the nuclear position. The specific rate for the substitution under irreversible conditions can be written $k = k_1 k_2/(k_{-1} + k_2)$ if the reaction proceeds as shown on p. 109 and the steady state method applies. (If a base B enters the second transition state, k_2 should be exchanged for $k_2[B]$.) In the particular compound used it could be expected that steric effects would increase the rate of step -1 relatively to that of step 2. Such an effect might be able to switch the ordinary $k_{-1} \ll k_2$ over to $k_{-1} \geq k_2$; that is, the ordinary gross specific rate would change from $k = k_1$ toward $k \approx k_1 k_2/k_{-1}$, and the isotope effect on k_2 would be reflected by k. In experiments with nitric acid in acetic acid plus acetic anhydride, however, the effect is still missing. (Compare, on the other hand, Zollinger's results below.)

Bromination. Some evidence for the absence of the hydrogen isotope effect in bromination was obtained by Melander (64), who used iodine as a catalyst. De la Mare, Dunn, and Harvey (27) measured the rates of bromination of benzene-d_6 and ordinary benzene with a solution of hypobromous and perchloric acids. Under these conditions the brominating agent is believed to be Br^+ or $BrOH_2^+$. The rates were determined in independent runs and found to be equal within a few per cent.

Alkylation. In a recent paper Bethell and Gold (8) report that the reaction

$$(C_6H_5)_2CH \cdot OOCCH_3 + C_6H_5OCH_3 \longrightarrow$$
$$\longrightarrow \quad p\text{-}CH_3O \cdot C_6H_4 \cdot CH(C_6H_5)_2 + CH_3COOH$$

proceeding in acetic acid solution containing some sulfuric acid, shows no deuterium isotope effect within a few per cent. The runs were independent, some with anisole-4-d and some with ordinary anisole. The rate-determining step here is probably the addition of the carbonium ion to the aromatic nucleus.

Discussion. Before more experimental evidence is presented, we shall briefly consider the fact that the addition step might actually proceed without a measurable isotope effect. Sometimes the rates have been found to be equal within a few per cent; this also excludes effects of the strength generally encountered in the case of secondary isotope effects.

As pointed out by Hammond (45), the absence of a measurable isotope effect is no more than an indication that the transition state of the over-all reaction is reached before the pertinent carbon-hydrogen bond has lost an appreciable part of its zero-point energy. It is not necessary to assume the existence of an intermediate, or, what amounts to the same, a minimum in the potential-energy curve. This is certainly true, but, if the reverse reaction is considered, this would mean that we can bring the hydrogen up to its normal position without gaining so much energy in σ-bond formation that this gain ever balances the loss due to stretching of the carbon-substituent bond. This does not seem entirely probable, and the intermediate mechanism is more likely. The preparative results of Olah and co-workers (71) prove that the intermediate might have considerable stability, and its real existence seems now generally accepted.

One must still explain, however, how isotopic substitution can have such a negligible influence on the rate of the addition step. In the transition state the initial sp^2 carbon-hydrogen bond must have been changed at least somewhat in the direction toward sp^3. Owing to the low frequency of the out-of-plane bending mode of aromatic carbon-hydrogen bonds (3c), it could be expected that the transition from sp^2 to sp^3 should increase the zero-point energy and consequently cause a secondary α-isotope effect with the heavy molecule reacting faster. On the other hand, as pointed out by Streitwieser et al. (93), hyperconjugation of the carbon-hydrogen bond removed from the ring plane with the p-orbitals of the other five carbon atoms will tend to decrease the zero-point energy, and the two effects probably cancel each other. It would certainly be of interest to try to increase the experimental accuracy still further in order to find a small isotope effect and to see which direction it has. Even the direction is unknown at present.

Some of the compounds used have been deuterated in all nuclear positions. It is certainly interesting, but not particularly astonishing, to find that deuteration in the non-reacting positions has a very slight influence on the reaction rate. Secondary isotope effects due to hyperconjugation with the π-electron system are excluded since the carbon-hydrogen bonds are situated in the ring plane and thus their orbitals are orthogonal to the carbon p-orbitals.

Sulfonation. By no means do all electrophilic aromatic substitutions proceed without a hydrogen isotope effect. Sulfonation has long been known (64) to proceed with a weak but finite isotope effect, the heavy compound reacting slower. The reaction has been investigated particularly by Berglund-Larsson (6), who sulfonated bromobenzene with oleum in nitrobenzene solution. The experiments were of the competitive type, and small amounts of bromobenzene-4-d or -4-t were used in mixture with ordinary bromobenzene. k_D/k_H was of the order of 0.7, and k_T/k_H of the order of 0.5, these values being mainly independent of the temperature in the interval 0° to 50°C. In a private communication Dr. J. C. D. Brand has reported the value $k_D/k_H = \frac{2}{3}$ at 25°C for the sulfonation of nitrobenzene-d_5 and nitrobenzene with oleum.

In the case of sulfonation the reaction scheme is somewhat complicated by the fact that the entering substituent is able to give rise to an acid-base equilibrium. The strength of ordinary sulfonic acids makes it probable that the intermediate should be written

also under highly acid conditions, because it contains a positive charge which has no counterpart in the ordinary acids. The negative charge not far from the hydrogen to be split off from the intermediate in the final step is characteristic for sulfonation and likely to slow down the proton expulsion. In this way the latter might have become partly rate-determining.

Azo Coupling. The transition from the first stage being rate-controlling to the second stage being rate-controlling and its manifestation in the isotope effect have been demonstrated elegantly by Zollinger (105) in his investigations of the azo coupling reactions. The substitution step of azo couplings in general proceeds without hydrogen isotope effect and without being catalyzed by bases. If, however, the molecule to be substituted is sufficiently sterically hindered, the rate of reaction -1 (cf. p. 109) might be sufficiently increased relative to the rate of reaction 2 to make the latter rate-determining. This results in catalysis by bases as well as an appreciable isotope effect. It is obvious that the gross specific rate,

$$k = \frac{k_1 k_2 [B]}{k_{-1} + k_2 [B]} = \frac{k_1}{\dfrac{k_{-1}}{k_2} \times \dfrac{1}{[B]} + 1}$$

has undergone a transition from $k_{-1} \ll k_2[B]$ to $k_{-1} \gg k_2[B]$. One of the most interesting achievements of Zollinger is that he has shown that the isotope effect might change with changing base concentration, $[B]$. Since neither k_1 nor k_{-1} should depend on the isotopic mass of the hydrogen, any measured effect arises from k_2. With proper choice of reactants it is possible to let $[B]$ decide the relative order of magnitude of the two terms in the denominator. This was indeed possible in the coupling between 4-chlorobenzenediazonium ion and 2-naphthol-6,8-disulfonic acid, which carried deuterium in position 1 in one set of separate runs. By means of variation of $[B]$ it was possible to separate out k_1 and the ratio k_{-1}/k_2 for each isotopic reactant. Hence it was also possible to compute $k_{2(D)}/k_{2(H)}$, which was found to be 0.156 ± 0.006 when the over-all isotope effect varied between this value and 0.28 at 10°C. This is a very good confirmation of the idea of a two-stage reaction in electrophilic aromatic substitution in general. Zollinger (106) also reports the observation of an isotope effect in the bromination of 2-naphthol-6,8-disulfonic acid, k_D/k_H being about 0.5, so the manifestation of steric hindrance is probably quite general, although Myhre's result (see above) failed to show such an influence in a probably sterically hindered nitration.

Binks and Ridd (15) have reported that the couplings of p-nitrobenzenediazonium ion with indole and indole-3-d at

0°C proceed with the same velocity within a few per cent, as determined in separate runs. This is, of course, a coupling subject to very little steric hindrance.

Iodination versus Azo Coupling. Grimison and Ridd (40) have recently reported the very interesting result that diazotized sulfanilic acid couples with glyoxaline, probably in the form of its conjugate base,

in position 2 with negligible isotope effect, while iodination of the same compound (also probably the conjugate base) initially attacks position 4; experiments with glyoxaline-2,4,5-d_3 gave k_D/k_H about 0.23 for both catalyzed and uncatalyzed iodination. The isotope effect also served to settle the point of initial attack of iodine; glyoxaline-4,5-d_2 gave the isotope effect, while glyoxaline-2-d gave almost none. (The iodination does not stop sharply at the monoderivative.) This observation of an isotope effect in iodination agrees with the somewhat earlier result of Grovenstein and Kilby (42), who found $k_D/k_H = 0.25$ at 25°C for the iodination of phenol-2,4,6-d_3 and ordinary phenol in separate runs in aqueous medium. It seems that the second reaction step is rate-determining in these iodinations but not in ordinary azo coupling.

It is certainly an interesting fact that the orientation of the electrophilic attack on the conjugate base of glyoxaline is different for two substitution reactions, one of which has the first and the other the second step rate-determining. This agrees nicely with the result of molecular-orbital calculations presented by Bassett and Brown (1). Reasonable assumptions about the magnitude of a certain quantum-mechanical parameter lead to the prediction that carbon atom 2 of the glyoxaline anion should carry a larger electronic charge than carbons 4 and 5. It seems natural that the rate of the first reaction step should be correlated with the charge distribution in the unperturbed reactant. The so-called atom localization energy, representing the difference in π-electron energy between the substitution intermediate and the reactant, becomes larger in position 2 than in position 4 or 5 on the same assumptions. It seems natural also that the rate of the second reaction step should be correlated with the stability of the intermediate, i.e., that the potential energy of the second transition state should run parallel with that of the intermediate. Thus, in azo coupling, the first transition state has the highest potential energy, and its magnitude seems to follow the unperturbed initial charge distribution. In iodination, the second transition state is at the highest energy level, and its height seems to follow that of the intermediate.

Cyclizations. A weak isotope effect has been observed by Bonner and Wilkins (18) in the cyclodehydration in aqueous sulfuric acid of 2-anilinopent-2-en-4-one and the same compound trideuterated in the aromatic positions 2, 4, and 6. The reaction proceeds probably in the conjugate acid:

Separate runs with the two isotopic compounds gave $k_D/k_H = 0.7$ at 25°C. It is not easy to draw any definite conclusions about the source of this weak effect. Bonner and Wilkins seem to favor a two-stage mechanism with the first stage being rate-determining and accompanied by the observed effect. This is of course not excluded, although it would generally seem more natural to assume an effect in the direction $k_D > k_H$, as explained above. The fact that the entering substituent is already joined to the aromatic ring might impose some particular steric conditions on the transition state in the present case.

Unusual steric conditions of the same kind certainly prevail also in the acid-catalyzed cyclizations of 2-carboxybiphenyl, which have been investigated by Denney and Klemchuk (28). The pertinent reaction is probably

One of the two positions toward which the attack is directed was completely deuterated, and the isotope effect was studied as an exclusively intramolecular competition, which will produce the two possible products in constant ratio all through the reaction. The values were of the order $k_D/k_H = 0.75$ to 0.88 at 1°C when the reaction was catalyzed by aqueous sulfuric acid of different concentrations; 0.68 at 25°C and 0.76 at 95°C when it was catalyzed by polyphosphoric acid; and 0.33 at 19°C with anhydrous hydrogen fluoride. Since the isotope effect seems to grow stronger with increasing acidity, it seems to be a likely explanation that both steps are partly rate-determining, the second one being of increasing kinetic significance, exactly as in the azo coupling investigated by Zollinger (105); cf. pp. 113 ff. A similar trend appears also in the results of Bonner and Wilkins, described above. More work has certainly to be carried out before the influence of steric effects arising from a bridge between the aromatic nucleus and the attacking electrophile can be ascertained.

Electrophilic Hydrogen Exchange. The theoretically simplest electrophilic aromatic substitution is the electrophilic hydrogen exchange, in which a hydrogen nucleus attached to an aromatic nucleus is replaced by another. If the two hydrogen nuclei are of different mass, the reaction involves an isotope exchange. Since there are three hydrogen isotopes, it is possible to study the isotope effect in the exchange, the two exchange reactions to be compared being conducted under quite comparable conditions. Thus, for instance, deuterium and tritium have been replaced by protium in the experiments of Olsson (69, 72), who studied the exchange in benzene and toluene with aqueous sulfuric acid as catalyst.

Owing to side reactions (sulfonation, oxidation) and the existence of several different exchange reactions and equilibria in the case of toluene, the initial rates were measured and the exchange was generally not followed until equilibrium had been established. According to pp. 63 ff., such measurements give the isotope effect in the forward reaction.

The reaction, for instance with benzene-d, might probably be written

with the specific rate of the forward reaction

$$k = \frac{k_1 k_2}{k_{-1} + k_2} = \frac{k_1}{\dfrac{k_{-1}}{k_2} + 1}$$

If a base takes part in the transition states, k_2 and k_{-1} should be replaced by $k_2[B]$ and $k_{-1}[B]$, respectively; this does not change the expression for the over-all k. It is, however, probably a better description of the transition state of the reaction in aqueous solution to exclude water from it, as pointed out by Melander and Myhre (68), who also show that the presence of π-complexes as intermediates, frequently referred to in the literature, is neither required nor proved by the kinetics. Since k_1 and k_{-1} are probably little affected by the isotopic mass of the hydrogen to be replaced, we could write for the replacements of tritium and deuterium, at least as a first approximation,

$$\frac{k_T}{k_D} = \frac{\dfrac{k_{-1}}{k_{2(D)}} + 1}{\dfrac{k_{-1}}{k_{2(T)}} + 1}$$

The attractive feature of this reaction is that it is completely symmetric in reactants and products except for the mass of the hydrogen. Hence k_{-1} and k_2 cannot be of different orders of magnitude in such a way that it is possible to talk of a "rate-determining" first or second reaction step. In fact, the ratio k_{-1}/k_2 has the nature of an intramolecular hydrogen isotope effect. Thus, for instance, $k_{-1}/k_{2(D)}$ is the ratio of the specific rates for protium and deuterium loss from the intermediate of the type $C_6H_6D^+$. If the intermediate behaves like a normal molecule, the magnitude of k_T/k_D can be approximately predicted. A common value for a simple k_D/k_H ratio at room temperature is $\frac{1}{4}$, and thus we could tentatively assign to $k_{-1}/k_{2(D)}$ the value 4. According to Eq. (2–13) the corresponding assignment of $k_{-1}/k_{2(T)}$ would be $(4)^{1.44} - 7.4$; hence $k_T/k_D = 5/8.4 = 0.60$. This has indeed been found by Olsson, the values of k_T/k_D for benzene and the different positions of toluene being in the range 0.52 to 0.65 at 25°C. All these determinations refer to the same medium. Recent results with benzene, obtained by Dr. Morley Russell, indicate that the isotope effect does not change appreciably with the acid concentration.

Olsson's results certainly do not prove the two-stage mechanism, because a direct one-stage replacement could also have the values $k_D/k_H = \frac{1}{4}$ and $k_T/k_H = 1/7.4$, making $k_T/k_D = 4/7.4 = 0.54$. Thus, reasonable assumptions give about the same prediction for either mechanism, and the prediction is entirely consistent with the experimental results. The reason why the mechanism has been discussed mainly in terms of the two-stage process is, of course, the analogy with those substitution reactions which show no measurable isotope effect, for instance nitration.

The value of the method depends on the information it seems to be able to give about the vibrational behavior of the intermediate. Much work has to be done before definite conclusions will be allowed. At present, however, it seems there is some correlation between the isotope effect and the absolute rate of the reaction, the slowest reaction being accompanied by the weakest isotope effect.

Decarbonylation. The results hitherto described have referred to replacements of isotopic hydrogen, this type of reaction being by far the best-known one in the electrophilic class of substitutions. The investigations of Schubert and Burkett (86) and of Schubert and Myhre (87) concerning the electrophilic decarbonylation of 2,4,6-trimethyl- and 2,4,6-triisopropylbenzaldehyde in aqueous sulfuric acid at 80°C are examples of how hydrogen isotope effects might be used in other ways. It is not possible to discuss all the interesting features of these experiments here, and only the bare outlines can be given. The mechanism of the reaction seems to be (HA$_i$ is a general acid, H_2SO_4 or H_3O^+)

in agreement with the general scheme of electrophilic aromatic substitution. The aldehyde is also tied up by conjugate acid formation at the oxygen atom, but this entity could probably be treated as non-reactive. The relative rates of the two steps are different under different conditions.

This scheme might be tested with isotopic hydrogen in the aldehyde group and in the medium. When R is equal to isopropyl and the concentration of the sulfuric acid is above 85%, the isotopic mass of the hydrogen in the aldehyde group has hardly any influence on the rate, k_D/k_H being in the range 0.93 to 0.99. In this case the first step is rate-determining. Under the same conditions with R equal to methyl there is a weak isotope effect, $k_D/k_H = 0.36$ to 0.56. This is probably due to less steric hindrance in reaction -1, making the first step partly reversible and the second partly rate-determining. It should be observed that the reactions -1 and 2 are analogous in the sense that they are both hydrogen abstractions, but, owing to the position of the hydrogen to be abstracted, reaction -1 is probably much more hindered than reaction 2. Thus changes in R will influence k_{-1i} and k_{2i} differently, and the ratio k_{-1i}/k_{2i} will be a function of R as well as of the nature of the basic species A_i. Thus with sulfuric acid in the concentration range 70 to 85%, the triisopropyl compound also shows a slight aldehyde-group isotope effect, k_D/k_H being about 0.8. In this medium, water is probably the main base, and it is less sensitive to steric effects than the bisulfate ion. Hence reaction -1 is no longer entirely negligible. The trimethyl compound shows an isotope effect of about the same strength as in the more concentrated acid.

When the hydrogen of the sulfuric acid is replaced by deuterium, the first step introduces isotope effects, and the second is not perceptibly influenced. In concentrated acid with R equal to isopropyl, we have already seen that the bulky R and the bisulfate ion probably introduce considerable steric hindrance in step -1, which is then slow, and according to the customary terminology the first step is rate-determining. This means that the transference of the hydrogen from the acid to the aromatic hydrogen should be rate-determining and probably cause an isotope effect. This agrees with the observation that the over-all reaction has $k_D/k_H = 0.40$ in 99 per cent sulfuric acid (when the difference in conjugate acid formation at oxygen between the isotopic media has been taken into account). For the methyl compound the corresponding value was about unity. At an acid concentration of about 70 per cent the isopropyl compound still gave a value of about 0.50, but the methyl compound gave about 2.2. In 59 per cent sulfuric acid, finally, the methyl compound gave the value 2.3. The latter limiting values are very easily explained by the assumption that there is full equilibrium in the first reaction step, and the two reaction rates are proportional to the concentration of this intermediate (no isotope effect of this kind in the second step). In the heavy medium the concentration of the intermediate relative to the concentration of free reactant should thus be 2.3 times larger than in the ordinary medium. The difference in the H_0 function, as measured from the acid-base equilibrium at the oxygen atom of the methyl compound, is 0.35, the heavy medium being the more acidic one. Since the antilogarithm of 0.35 is 2.2, the agreement is excellent.

The investigation described shows that "medium isotope effects" might give rate ratios $k_D/k_H \gtreqless 1$, depending on the mechanism. Such effects will be very briefly discussed at the end of this chapter.

NUCLEOPHILIC AROMATIC SUBSTITUTION

Owing to the fact that the hydrogen anion is much less stable than the cation under ordinary chemical conditions, replacements of aromatic hydrogen by nucleophilic agents are much less energetically favored than the corresponding electrophilic reactions. Nucleophilic aromatic substitution of hydrogen has much less importance than, for instance, nitrations and halogenations. Nucleophilic replacement of groups capable of existing as anions or some other entity with a lone electron pair is much easier and of greater preparative importance.

"Element Effect." A very interesting investigation in this field has been carried out by Bunnett, Garbisch, and Pruitt (24). Since this work does not deal at all with isotopes, it could possibly be claimed that it deserves no space in this book, but owing to its close relationship to the isotope methods it should certainly be mentioned in connection with the isotope work on electrophilic aromatic substitution. The nucleophilic substitution of nine 1-substituted 2,4-dinitrobenzenes with piperidine in methanol, forming 2,4-dinitrophenylpiperidine, has been studied. The substituents were F, NO_2, $OSO_2C_6H_4CH_3$ (p), SOC_6H_5, Br, Cl, $SO_2C_6H_5$, $OC_6H_4NO_2$ (p), and I. Except for the first three, the compounds of which reacted faster, all showed the same second-order specific rate at 0°C within a factor 4.7, and, if iodine also is excluded, a factor of 1.6. Measurements at different temperatures showed that the analogy was valid also for the energies and entropies of activation.

The fact that 6 (with the exception of iodine 5) different groups, representing 5 (4) different elements attached to carbon, are replaced at very nearly the same rate is at first sight very striking. This absence of an "element effect" is, however, closely analogous to the absence of an isotope effect of hydrogen in electrophilic aromatic substitution. Since it is known that the rate of cleavage of bonds between carbon and other elements varies strongly from element to element, the constant rate here can hardly be explained in any other way than by the assumption that the bond is not cleaved in the rate-determining step. Thus an ordinary S_N2 reaction seems to be excluded. As in the electrophilic case, the addition of the attacking agent is probably rate-controlling. The fact that the three first reactions were much faster than the others does not invalidate the above conclusion. Nothing excludes the possibility that some substituents should be able to exercise a particular, accelerating influence on the addition step.

RADICAL AROMATIC SUBSTITUTION

The evidence from measurements of the hydrogen isotope effect in radical substitution of aromatic nuclei is somewhat conflicting. The individual investigations will not be described here. The reaction has recently been discussed by Eliel, Welvart, and Wilen (33), who present evidence of heavy isotope enrichment in the products of several substitutions of this class and point to the fact that a branched second step in the general two-step mechanism will introduce such an isotope effect without the first step being appreciably reversible. The kinetic situation is of considerable interest for isotope effect studies and will therefore be discussed here.

The reaction scheme is

where $X\cdot$ denotes some species capable of abstracting the hydrogen atom from the intermediate. In the present case there are five electrons in the π-electron system of the intermediate, which carries no net charge. The steady state condition here involves (square brackets denote concentration):

$$k_1[AH][R] = (k_{-1} + k_2[X] + k_2')[AHR]$$

giving

Rate of formation of AR =

$$= k_2[AHR][X] = \frac{k_1 k_2}{k_{-1} + k_2[X] + k_2'}[AH][R][X] =$$

$$= \frac{k_1}{\dfrac{k_{-1} + k_2'}{k_2[X]} + 1}[AH][R]$$

If other chemical species take part in the formation of the side products, their concentration is considered included in the symbol k_2' above and in the following. Of the specific rates occurring above, k_2 will certainly be sensitive to the mass of the hydrogen and k_1 and k_{-1} probably not. Thus, if under some conditions the product AR is formed without significant isotope fractionation, we conclude $(k_{-1} + k_2') \ll k_2[X]$. On the other hand, if there is a measurable isotope fractionation in the formation of AR, $(k_{-1} + k_2')$ will be of at least the same order of magnitude as $k_2[X]$. The latter condition does not necessarily imply that k_{-1} need be large or even different from zero. An appreciable side reaction, i.e., a sufficiently large k_2', is reason enough to cause an isotope effect in the formation of AR.

Eliel, Welvart, and Wilen point out that the results of Convery and Price (26) make a reversible first step less probable.

The latter writers arylated *m*-dinitrobenzene containing tracer amounts of 1,3-dinitrobenzene-4-*t* with free phenyl radicals and found that the specific tritium content of the *m*-dinitrobenzene is not appreciably changed even after over 75 per cent conversion. Since the total rate of consumption of AH can be written

Total rate of consumption of AH $= (k_2[X] + k_2')[AHR] =$

$$= \frac{k_1(k_2[X] + k_2')}{k_{-1} + k_2[X] + k_2'}[AH][R] = \frac{k_1}{\dfrac{k_{-1}}{k_2[X] + k_2'} + 1}[AH][R]$$

and k_2 is likely to be sensitive to the mass of the hydrogen, and, moreover, k_2' is probably considerably smaller than $k_2[X]$ in this particular case, the natural explanation of the observation seems to be $k_{-1} \ll (k_2[X] + k_2')$, i.e., a mainly irreversible first step. For other reactions, then, it might be inferred that side-product formation is the most likely explanation of heavy-hydrogen enrichment in the product AR. It is obvious that the observed isotope effect should also be sensitive to the composition of the reaction mixture, because $k_2[X]$ and possibly k_2' are sensitive. Such a dependence was also observed by Eliel *et al.*

DEUTERIUM SOLVENT EFFECTS

Since deuterium is available in macroscopic quantities, it might be used for more or less complete replacement of protium even in solvents, particularly water, and reagents, such as acids, which are in rapid protolytic exchange with the solvent. Many reactions susceptible to acid-base catalysis have been studied in a heavy-solvent system or in a solvent containing an appreciable fraction of deuterium. This type of experiment will not be discussed in detail here. References to such work are found in Bell's well-known monograph, *Acid-Base Catalysis* (2), and in Wiberg's review (101). A very recent paper on solvent isotope effects in mixtures of heavy and ordinary water has been published by Purlee (74).

From the results of Schubert, Burkett, and Myhre (86, 87) discussed on pp. 120 ff., it is evident that the solvent isotope effect might take either direction, $k_D/k_H \gtreqless 1$, depending on the mechanism. If a hydrogen transfer from the solvent or some acid solute is rate-determining, a normal isotope effect will generally arise in this particular reaction step. If, on the other hand, there is a rapid pre-equilibrium, in which the reactant is converted to its conjugate acid, the latter intermediate will generally be present in higher equilibrium concentration in the heavy medium, because the dissociation of weak acids is generally smaller in the heavy system (63, 84). If this intermediate gives the product in a second, rate-controlling step, and the latter step gives rise to no further isotope effect, the reaction will be fastest in the heavy system. Finally, if the second step, too, has an isotope effect, the over-all isotope effect will be the result of these two competing effects. It is hardly astonishing, then, that the over-all isotope effects observed in different reactions cover a range from a strong effect in one direction to a strong effect in the other direction.

In the quantitative comparison of reactions in the heavy medium with the same reactions in the ordinary medium, great care should be taken, and the fact that the media are actually different should be kept in mind, because, in principle, all kinds of medium effects could be different.

7

Carbon Isotope Effects in Single Reaction Steps

Of the elements other than hydrogen for which kinetic isotope effect measurements have been undertaken, carbon is by far the most important. The stable heavy isotope C^{13}, the mean natural abundance of which is about 1 per cent, and the radioactive isotope C^{14} offer rich experimental possibilities, and with modern techniques it is possible to measure fairly accurately the isotope effects (generally of the order of a few per cent) which arise.

There are some features which make the theoretical treatment of isotope effects of carbon and similar elements more difficult than that of hydrogen isotope effects. In ordinary organic compounds carbon is approximately at least as heavy as its neighbors. This fact makes it impossible to ascribe a vibration frequency "of its own" to carbon, as it is possible to do for hydrogen. The vibrational frequencies are thus often less sensitive to mass changes in carbon, and this, together with the fact that the relative mass difference between carbon isotopes (maximum 2 units in 12) is very much less than that between hydrogen isotopes (1 or 2 units in 1), makes the effects of much smaller magnitude. As pointed out in Chapter 2, it is, in general, no longer possible to ascribe the effect to the zero-point-energy factor alone, not even as a rough approximation. In order to avoid estimations of different magnitudes, which are to balance one another in the final result, it is frequently better to use Bigeleisen's expressions, which concentrate all these effects in a function of the vibrational frequencies.

Another difficulty in the theoretical handling of the isotope effects of carbon arises from the fact that the isotopic atom is attached to several ligands (with the exception of carbon monoxide). Thus a molecule isotopic in carbon (or its closest neighbors in the periodic system) will contain several "isotopic bonds," and the breaking of one of them is not without influence on the characteristics of the others.

On the other hand, there is one feature which sometimes tends to make the computations easier. From the point of view of carbon, the transition state formation in several organic reactions seems to be much more a process of bond breaking than is the case with hydrogen in, for instance, hydrogen abstraction reactions. This is a consequence of the more protected position of carbon, but it is by no means a rule. Thus, for instance, all truly bimolecular displacements at carbon do not belong to this class. Among the reactions which have hitherto been studied, however, a non-bonded model is generally a fairly good approximation of the transition state.

DECARBONYLATION OF FORMIC ACID

The decarbonylation of formic acid could be taken as a good example of a reaction the carbon isotope effect of which has been determined accurately and the theoretical treatment of which might be carried through with very simple models.

Ropp, Weinberger, and Neville (83) used formic acid containing tracer amounts of formic acid-C^{14}, which was dehydrated in an excess of 95% sulfuric acid. The instantaneous specific activity of the evolved carbon monoxide and its volume were followed as functions of time. The reaction was shown to be of first order. Since the macroscopically observable substance consists almost exclusively of ordinary formic acid, the first-order law obviously holds for each isotopic kind separately.

If we denote the amount of C^{14} acid by a_{14} and that of the C^{12} acid by a_{12} (initial amounts a_{14}^0 and a_{12}^0), it is possible to write

$$a_{14} = a_{14}^0 e^{-k_{14}t} \qquad a_{12} = a_{12}^0 e^{-k_{12}t}$$

The instantaneous specific activity of the evolved carbon monoxide is proportional to the fraction of $C^{14}O_2$, given by the expression

$$y = \frac{-da_{14}}{-d(a_{14} + a_{12})} \approx \frac{da_{14}}{da_{12}} = \frac{k_{14}a_{14}^0}{k_{12}a_{12}^0} e^{(k_{12}-k_{14})t}$$

That means that the instantaneous specific activity of the carbon monoxide also shows first-order behavior:

$$\frac{d \ln y}{dt} - k_{12} - k_{14}$$

Since k_{12} is available for direct determination, it is possible to obtain k_{14}/k_{12} from the latter and from the slope of the linear growth of the logarithm of the instantaneous specific activity. The value obtained at 0°C was $k_{14}/k_{12} = 0.88_9$ and, at 24.75°C, 0.91_4.

It seems probable that the slow step of the reaction is the fission of the bond between carbon and the oxygen of the protonated hydroxyl group (cf. p. 157):

$$\text{H}-\text{C}\overset{\displaystyle O}{\underset{\displaystyle OH_2^+}{\diagdown}} \longrightarrow [\text{HCO}]^+ + \text{OH}_2$$

Eyring and Cagle (35) have used the very crude model of the dissociation of a hypothetical diatomic molecule C—O for a computation of the isotope effect (Eq. 2–7). According to Bonner and Hofstadter (16), the stretching frequency of the carbon-oxygen single bond in ordinary formic acid (i.e., a C^{12}—O^{16} bond) is 1093 cm^{-1}. For a corresponding diatomic molecule C^{14}—O^{16}, the frequency is calculated with the aid of the reduced mass ratio:

$$\bar{\nu}_{14} = \bar{\nu}_{12}\left(\frac{\dfrac{1}{14} + \dfrac{1}{16}}{\dfrac{1}{12} + \dfrac{1}{16}}\right)^{1/2} = 1047 \text{ cm}^{-1}$$

where $\bar{\nu}$ denotes wave number, and the mass number of the particular carbon atom is used as index.

Equation (2–7) is almost as good as Eq. (2–6) in this case, because for a unimolecular decomposition the molecular mass term of the latter is identically equal to unity, and the term for the moments of inertia cannot differ much from unity. (For a real diatomic molecule the latter term also would be identically equal to unity; cf. pp. 151–152.)

Since the isotopic substitution does not change the molecular symmetry, the symmetry numbers cancel and Eq. (2–7) gives, at 0°C,

$$\frac{k_{14}}{k_{12}} = \frac{\sinh \dfrac{hc \times 1047}{2k \times 273}}{\sinh \dfrac{hc \times 1093}{2k \times 273}} = 0.885$$

Owing to the fairly high frequencies, Eq. (2–10) could have been used as well. The result becomes 0.886. At 25°C, Eqs. (2–7) and (2–10) give 0.894 and 0.895, respectively.

The agreement with the experimental data is surprisingly good in view of the crude model. It is evident that the experimental temperature coefficient of the isotope effect is larger than the one required by the theory, but, on the other hand, the discrepancy is certainly not considerably beyond the experimental uncertainty.

It is of interest to compare the above predictions with those obtained by means of Bigeleisen's heavy-atom approximations Eqs. (2–18) and (2–19), because the approximations are introduced somewhat differently.

The temperature-independent factor $\nu_{L(14)}^{\ddagger}/\nu_{L(12)}^{\ddagger}$ could be calculated from the atomic masses at the ends of the bond to be broken or from the masses of the two fragments which are separated. The first method (Eq. 2–20) gives

$$\frac{\nu_{L(14)}^{\ddagger}}{\nu_{L(12)}^{\ddagger}} = \left(\frac{\dfrac{1}{14} + \dfrac{1}{16}}{\dfrac{1}{12} + \dfrac{1}{16}}\right)^{\frac{1}{2}} = 0.9583$$

and the second (Eq. 2–21)

$$\frac{\nu_{L\,(14)}^{\ddagger}}{\nu_{L\,(12)}^{\ddagger}} = \left(\frac{\frac{1}{31} + \frac{1}{18}}{\frac{1}{29} + \frac{1}{18}}\right)^{\frac{1}{2}} = 0.9876$$

Using Eq. (2–18), we make the approximation that all vibrations but the stretching of the pertinent carbon-oxygen bond have the same frequencies in the transition state as in the initial molecule. Since all contributions which are equal for the reactant and the transition state disappear by cancellation, as do the symmetry numbers, the remaining expression becomes

$$\ln \frac{k_{14}}{k_{12}} = \ln \frac{\nu_{L\,(14)}^{\ddagger}}{\nu_{L\,(12)}^{\ddagger}} + G(u_k)\,\Delta u_k$$

where u_k refers to the carbon-oxygen stretching mode in question. Using the above values, 1093 and 1047 cm^{-1}, we find at 0°C

$$u_k = u_{k\,(12)} = \frac{hc \times 1093}{k \times 273} = 5.761$$

$$\Delta u_k = -\frac{hc \times 46}{k \times 273} = -0.242$$

From a table (12, 32) we find

$$G(5.761) = 0.330$$

and, using the temperature-independent factor according to the simple diatomic model,

$$\frac{k_{14}}{k_{12}} = 0.9583 \times \exp\,(-0.330 \times 0.242) = 0.885$$

If the simpler Eq. (2–19) is used, we obtain

$$\frac{k_{14}}{k_{12}} = 0.9583(1 - 0.330 \times 0.242) = 0.882$$

At 25°C, Eq. (2–19) gives $k_{14}/k_{12} = 0.891$. According to the discussion on p. 35 it is obvious that the present assumptions should give the same prediction when used in Eyring-

Cagle's Eq. (2–7) as when used in Bigeleisen's expressions. The slight difference that is still found is caused by the approximations inherent in Eqs. (2–18) and (2–19).

Hitherto we have not used the temperature-independent factor calculated according to Eq. (2–21). With the same frequencies as above, the predictions of such a calculation according to Eq. (2–19) are $k_{14}/k_{12} = 0.909$ at 0°C and 0.918 at 25°C. The agreement with the experimental values, 0.88_9 and 0.91_4, is about as good as with the previous temperature-independent factor.

DECARBOXYLATION OF MALONIC ACID

The decarboxylation of dibasic carboxylic acids is perhaps the type of reaction whose carbon isotope effect has been most thoroughly studied by different research workers. The following is not an attempt to review this work; it will rather try to exemplify the problems involved.

Malonic acid, $CH_2(COOH)_2$, may form two different isotopic isomers containing one heavy carbon atom. The one which contains the heavy atom in the methylene group could expel carbon dioxide in two identical ways, the molecule being isotopically symmetric. The unsymmetric isomer could react in two different ways, the heavy carbon being contained either in the carbon dioxide or in the acetic acid.

An experimental system using C^{14} will generally consist of tracer amounts of one of the two isomeric malonic acids, containing one heavy carbon atom, together with a macroscopic bulk of ordinary malonic acid. There will, of course, be a slight amount of natural C^{13}, both in the C^{14} molecules and in the ordinary ones, but, owing to the low natural ratio C^{13}/C^{12}, the corresponding molecules could be neglected and the sample considered to be made up exclusively of C^{14} and C^{12} atoms.

A natural malonic acid sample, which might also be used for isotope effect studies, will, of course, contain symmetric as well as unsymmetric C^{13} molecules and still smaller amounts of molecules containing two and even three C^{13} atoms. It is generally possible to neglect the molecules containing more than one C^{13} atom, owing to the low natural abundance.

The reaction scheme of a sample containing both isomers of mono-heavy malonic acid together with the ordinary compound is illustrated below for the case of C^{14} (the mass number of carbon is assumed to be 12 if not indicated):

$$CH_2 \underset{COOH}{\overset{COOH}{\Big\langle}} \xrightarrow{k_1} CH_3COOH + CO_2$$

$$C^{14}H_2 \underset{COOH}{\overset{COOH}{\Big\langle}} \xrightarrow{k_2} C^{14}H_3COOH + CO_2$$

$$CH_2 \underset{COOH}{\overset{C^{14}OOH}{\Big\langle}} \begin{cases} \xrightarrow{k_3'} CH_3COOH + C^{14}O_2 \\ \xrightarrow{k_3''} CH_3C^{14}OOH + CO_2 \end{cases}$$

Quite generally, k_1 and k_2 will of course be roughly equal to twice each of the k_3. Particularly, it could be expected that k_3'' would be almost exactly half as large as k_1, both reactions involving the cleavage of a pure C^{12}—C^{12} bond. In the following the symmetry factors will be automatically introduced via the symmetry numbers.

Some simple conclusions can be drawn without making detailed kinetic calculations. Once the reaction is unimolecular in malonic acid (and this seems certain), the unsymmetric molecular species will steadily produce heavy carbon dioxide and heavy acetic acid in the molecular ratio k_3'/k_3''. If no other kind of heavy molecule is present in the system, the absolute amounts of $C^{14}O_2$ and $CH_3C^{14}OOH$ formed in the reaction, instantaneously or cumulatively, will immediately give k_3'/k_3''. This holds irrespective of the relative amount of ordinary malonic acid present, i.e., irrespective of whether the heavy molecule is present in macroscopic or tracer amounts. In the latter case the bulk of reaction products will have come almost exclusively from the ordinary molecule, and acetic acid and carbon dioxide are steadily produced in equimolecular amounts. Thus the ratio of the molar activities of both product compounds also will give k_3'/k_3''.

C^{14} **Experiments.** Roe and Hellmann (79), in the way described above, determined the intramolecular isotope effect $k_3'/k_3'' = 0.94 \pm 0.02$ for malonic acid at 153° to 154°C. In these experiments the samples were decomposed quantitatively.

Ropp and Raaen (82) used this intramolecular isotope effect in another investigation of the rate-constant ratios involved in the general system above. The reaction, which occurs with molten malonic acid at 154°C without any other reagents being present, proceeds according to the first-order rate law.

In one set of experiments Ropp and Raaen followed the specific activity of the remaining part of a mixture of tracer amounts of malonic-1-C^{14} acid with a bulk of ordinary malonic acid. If the remaining amounts of the two isotopic compounds are denoted by a_3 and a_1, respectively, and the initial amounts were a_3^0 and a_1^0, we obtain

$$a_1 = a_1^0 e^{-k_1 t} \qquad a_3 = a_3^0 e^{-(k_3' + k_3'')t}$$

The specific activity of the remaining malonic acid will be proportional to the fraction malonic-1-C^{14} acid:

$$z = \frac{a_3}{a_1 + a_3} \approx \frac{a_3}{a_1} = \frac{a_3^0}{a_1^0} e^{(k_1 - k_3' - k_3'')t}$$

and the slope of its logarithmic plot versus time is equal to that of z,

$$\frac{d \ln z}{dt} = k_1 - k_3' - k_3'' = k_1 - 2.06 k_3'$$

the last equality being obtained by means of the result of Roe and Hellmann, referred to above. The first-order rate constant for ordinary acid was measured separately, and, from this value and the slope, k_3'/k_1 or, better, $2k_3'/k_1$ could be determined. The value was about $2k_3'/k_1 = 0.937$.

In another set of experiments the instantaneous specific activity of the evolved carbon dioxide was followed. It is proportional to the fraction $C^{14}O_2$ in the carbon dioxide, given by

$$y = \frac{-\dfrac{k_3'}{k_3' + k_3''} da_3}{-d\left(a_1 + \dfrac{k_3''}{k_3' + k_3''} a_3\right)} \approx \frac{k_3'}{k_3' + k_3''} \times \frac{da_3}{da_1} =$$

$$= \frac{k_3'}{k_1} \times \frac{a_3^0}{a_1^0} e^{(k_1 - k_3' - k_3'')t}$$

or

$$\frac{d \ln y}{dt} = k_1 - k_3' - k_3'' = k_1 - 2.06 k_3'$$

Thus the slope of the curve representing the logarithm of the instantaneous carbon dioxide specific activity should be the same as that of the curve mentioned above. This was also the case within the experimental accuracy, two experiments giving $2k_3'/k_1 = 0.940$ and 0.939.

Ropp and Raaen also decomposed a sample containing tracer amounts of malonic-2-C^{14} acid (82) and followed the specific activity of the remaining mixture. In this case we have, with analogous symbols,

$$a_1 = a_1^0 e^{-k_1 t} \qquad a_2 = a_2^0 e^{-k_2 t}$$

The specific activity of the remaining sample is proportional to

$$z' = \frac{a_2}{a_1 + a_2} \approx \frac{a_2}{a_1} = \frac{a_2^0}{a_1^0} e^{(k_1 - k_2)t}$$

and as before the slope of the logarithmic plot was studied,

$$\frac{d \ln z'}{dt} = k_1 - k_2$$

From the separately determined k_1 it is easy to compute k_2/k_1. The result was $k_2/k_1 = 0.929$, still measured at 154°C.

The last-mentioned value, $k_2/k_1 = 0.929$, together with the average $2k_3'/k_1 = 0.939$ allows a comparison of $2k_3'$ and k_2, giving $2k_3'/k_2 = 1.01$. The difference between $2k_3'$ and k_2 in evidently very small. Using the results of Roe and Hellman n once more, k_1 and k_3'' may be compared. Thus $2k_3''/k_1 = 1.00$. Also, these constants seem to be fairly equal to one another (however, see p. 142).

C^{13} Experiments. In this connection the early results of Bigeleisen and Friedman (11) should be mentioned. These workers decarboxylated ordinary malonic acid and determined the C^{13}/C^{12} isotope effects from the isotopic composition of the evolved carbon dioxide. As mentioned before, ordinary carbon contains about 1% of C^{13}. Bigeleisen and Friedman assumed that the heavy carbon was statistically distributed over the two different positions of the malonic acid. If α is the average fraction of C^{13} in ordinary carbon, it is easily shown that the initial amounts of the different isotopic species will be in the following ratio (indexing corresponding to that of the specific rates on p. 133):

$$a_1^0 : a_2^0 : a_3^0 = (1 - 3\alpha) : \alpha : 2\alpha$$

(Since the abundance of C^{13} is low, molecules with several heavy carbons need not be accounted for.) Since the decarboxylation is a first-order reaction, it is possible to write down the complete expression for the isotope ratio of the carbon dioxide as a function of time. This function is fairly unwieldy, and Bigeleisen and Friedman chose to use the initial isotope ratio and that of the total carbon dioxide after the reaction was complete. The expressions for these ratios are easy to find from the first-order behavior and a direct consideration of the scheme on p. 133. We use the same symbols, but it should be observed that those of the heavy molecules now refer to C^{13}.

$$\left(\frac{\text{amount of } C^{13}O_2}{\text{amount of } C^{12}O_2}\right)_{\text{initial}} = \frac{k_3' a_3^0}{k_1 a_1^0 + k_2 a_2^0 + k_3'' a_3^0} =$$

$$= \frac{k_3' \times 2\alpha}{k_1(1 - 3\alpha) + k_2\alpha + k_3'' \times 2\alpha} \approx \frac{2\alpha k_3'}{(1 - \alpha)k_1}$$

In the last expression use has been made of the approximations $\alpha k_2 \approx \alpha k_1$ and $2\alpha k_3'' \approx \alpha k_1$, which is allowed because of the small value of α when these magnitudes stand beside k_1. In a similar way we find

$$\left(\frac{\text{amount of } C^{13}O_2}{\text{amount of } C^{12}O_2}\right)_{\text{total, final}} = \frac{\dfrac{k_2'}{k_3' + k_3''} a_3^0}{a_1^0 + a_2^0 + \dfrac{k_3''}{k_3' + k_3''} a_3^0} =$$

$$= \frac{k_3' \times 2\alpha}{(k_3' + k_3'')(1 - 2\alpha) + k_3'' \times 2\alpha} =$$

$$= \frac{2\alpha k_3'}{k_3' + k_3'' - 2\alpha k_3'} \approx \frac{2\alpha k_3'}{(1 - \alpha)(k_3' + k_3'')}$$

where the last expression is obtained from the approximation that $\alpha k_3' \approx \alpha k_3''$ at the side of the much larger $k_3' + k_3''$. From measurements of the two isotope ratios and of α it is thus possible to compute $2k_3'/k_1$ and k_3'/k_3''. At 138°C the figures were $2k_3'/k_1 = 0.964$ and $k_2'/k_3'' = 0.981$ with an uncertainty of a few units in the last figure. These values for C^{13}/C^{12} compare very well with those for C^{14}/C^{12} given above, because the deviations from unity should be in the ratio 1:2 if observed at the same temperature.

That the two carbon isotope effects C^{13}/C^{12} and C^{14}/C^{12} should be related in the way mentioned above is evident from Bigeleisen's treatment, for instance Eq. (2–19). What will change in this expression are the Δu's and one of the $\nu_L^{\dagger}$'s. The isotope effects are a result of the mass difference between 13 and 12 in one case and between 14 and 12 in the other. Since Δu as well as $\nu_L^{\dagger}$ are slowly varying functions of the masses, the isotope effect could certainly be considered a linear function in this narrow interval. Bigeleisen and Wolfsberg (14) have made a comparison of these two carbon isotope effects for a number of reactions and have arrived at the conclusion that their relative magnitudes essentially agree with this theoretical prediction.

Theoretical Predictions. We are now to compare the experimental isotope effects with the detailed predictions obtainable from theory. All values given below will be for C^{14}/C^{12}. Several models for the transition state have been chosen, and more or less detailed models for the initial molecule have been tried. Of course, there is no possibility of using the complete vibrational pattern of either, but simplifying assumptions have to be made.

First we shall use the crudest picture, the diatomic C—C molecule which dissociates, and treat it according to Eyring and Cagle, Eq. (2–7). An approximate value of the stretching frequency of a bond like the one in question is 900 cm^{-1} according to the general table given by Herzberg (48a). As on p. 129, $\bar{\nu}_{12,14}$ is calculated from $\bar{\nu}_{12,12}$ and the respective reduced masses of the diatomic molecules. The result is $\bar{\nu}_{12,14} = 867$ cm^{-1}. From these frequencies we calculate for the temperature 154°C, neglecting symmetry factors, $k_{12,14}/k_{12,12} = 0.940$. The still simpler exponential of Eq. (2–10), accounting only for the zero-point-energy effect, gives 0.946.

Hitherto the isotope effect has not been related to the particular isomeric molecules. This will now be done as an illustration of how the symmetry numbers enter the calculation.

In the case of k_3' and k_1, the symmetry numbers are 1 and 2, respectively, for the reactants, and 1 for both of the transition states. Use of Eq. (2–7) thus gives

$$\frac{k_3'}{k_1} \times \frac{2}{1} \times \frac{1}{1} = \frac{k_{12,14}}{k_{12,12}}$$

or $2k_3'/k_1 = 0.940$.

With k_2 and k_1 the symmetry numbers of both reactants are 2 and those of both transition states 1, thus

$$\frac{k_2}{k_1} \times \frac{2}{2} \times \frac{1}{1} = \frac{k_{12,14}}{k_{12,12}}$$

or $k_2/k_1 = 0.940$.

With k_3' and k_3'' the reactant is common, and the transition states both have the symmetry number 1:

$$\frac{k_3'}{k_3''} \times \frac{s_3}{s_3} \times \frac{1}{1} = \frac{k_{12,14}}{k_{12,12}}$$

or $k_3'/k_3'' = 0.940$.

Similarly we obtain $2k_3'/k_2 = 1$ and $2k_3''/k_1 = 1$ according to our simple model, because in these cases the rate ratios are taken between two C^{12}—C^{14} or two C^{12}—C^{12} bond ruptures, respectively.

As pointed out in Chapter 2, Bigeleisen's treatment will give the same isotope effect as that of Eyring and Cagle when applied to this simple model, provided the Slater coordinate is used for the temperature-independent factor, i.e., Eq. (2–20). The single vibration mode which is not assumed to cancel between the transition state and the reactant gives a single $G(u_k) \Delta u_k$ term.

It might, however, be interesting to study how the cancellation works when (in principle) all vibrational modes of these fairly complex molecules are taken into account. Some modes might be assumed to be non-isotopic. They will give $\Delta u = 0$ or $\Delta u^\ddagger = 0$, and the corresponding terms vanish from the sums of Eq. (2–19). Other modes give rise to finite Δu or $\Delta u^\ddagger$, but, if the frequencies are not changed on passing into the transition state, there will be a corresponding pair-wise matching of terms between the two sums of Eq. (2–19). In simple treatments we talk about bond vibrations as if they were the normal modes of vibration, and then we are generally left simply with the stretching mode of the cleaving bond in the reactant, which corresponds to the reaction coordinate in the transition state and is not included in its sum. In the case of intramolecular isotope effects, however, things are somewhat different, as will be evident from a discussion of k_3'/k_3''.

In terms of the simple diatomic molecule occurring in the application of the formula of Eyring and Cagle, the reactant is simply treated as two different molecules, C^{12}—C^{14} and C^{12}—C^{12}, and the molecule as a whole need be considered only to find the symmetry numbers; cf. above.

Of course, the same kind of reasoning could be translated into a simplified Bigeleisen formalism, but this would be somewhat artificial. Since the actual reactant is one and the same in both isotopic reactions, all $\Delta u_i \equiv 0$. Equation (2–19) could therefore be written in the following way for any intramolecular isotope effect:

$$\frac{k_1}{k_2} \times \frac{s_1^\ddagger}{s_2^\ddagger} = \frac{\nu_{L(1)}^\ddagger}{\nu_{L(2)}^\ddagger}\left(1 - \sum_i^{3n^\ddagger-7} G(u_i^\ddagger)\,\Delta u_i^\ddagger\right)$$

Consequently, only the properties of the transition states concerned are decisive for the isotope effect. (Of course, the same conclusion is reached by making the proper cancellations in the primary expression of Eq. (2–5).)

When the two transition states

are compared, it is evident that they are isotopic in two positions. The carbon-carbon bond which is not broken corresponds to a non-vanishing term in the transition state sum. If the ratio k_3'/k_3'' is taken, $\Delta u^{\ddagger}$ should be the difference in $u^{\ddagger}$ between the first and the second of the two transition states above. $\Delta u^{\ddagger}$ is therefore positive.

The result is the same as if two reacting isotopic diatomic molecules, C^{12}—C^{14} and C^{12}—C^{12}, and two free-atom transition states had been considered. The single term was ascribed to the reactants, and the opposite sign in front of the summation symbol (cf. Eq. 2–19) was cancelled by the negative sign of Δu.

Several modifications of the simplest Bigeleisen-Eyring-Cagle treatment have been made, most of which are based on a more elaborate model of the initial molecule and the transition state.

The temperature-independent factor in Bigeleisen's expressions might, of course, also be calculated from the masses of the molecular fragments to be separated in the reaction. In the case of k_2/k_1 the factor will be

$$\frac{\nu_{L\,(2)}{}^{\ddagger}}{\nu_{L\,(1)}{}^{\ddagger}} = \left(\frac{1/45 + 1/61}{1/45 + 1/59}\right)^{\frac{1}{2}}$$

according to Eq. (2–21), giving with the above assumptions about the vibrational pattern $k_2/k_1 = 0.968$. In the same way, $2k_3'/k_1 = 0.964$ and $k_3'/k_3'' = 0.970$. Owing to this more elaborate computation of the reduced masses along the decomposition coordinate, $2k_3'/k_2$ and $2k_3''/k_1$ will no longer be equal to unity but to the square root of the reduced mass ratio concerned. Thus $2k_3'/k_2 = 0.995$ and $2k_3''/k_1 = 0.993$. The results are hardly better than those obtained with Slater's coordinate.

Bigeleisen (9) has calculated limiting values of $2k_3'/k_1$ and $2k_3''/k_1$ by taking the three skeletal modes of vibration of the reactant into consideration, and assuming that both carbon-carbon bonds are loosened in the transition state. This means that all three skeletal vibrations were assigned zero frequency in the transition state. All other frequencies were assumed to be the same in the reactant as in the transition state. Thus there are three terms of the type $G(u) \Delta u$ and none of the type $G(u^\ddagger) \Delta u^\ddagger$ in Eq. (2–19). The Slater coordinate was used. The results were $2k_3'/k_1 = 0.961$ and $2k_3''/k_1 = 0.997$ at 425 °K for C^{14}/C^{12}. To be a limiting value for the strength of the isotope effect, the former seems too weak.

According to this model of the transition state, the intramolecular isotope effect k_3'/k_3'' will be independent of temperature and will depend only on the reduced masses, because the transition state as well as the reactant are common to both reactions, and hence all $\Delta u^\ddagger$ and Δu are equal to zero. The result is then, according to Slater's method, simply $k_3'/k_3'' = 0.964$ at all temperatures.

Pitzer (73) has used the following model for a carboxylic acid:

$$R\!\!-\!\!C\overset{\displaystyle OH}{\underset{\displaystyle O}{\diagup}}$$

where R and OH are treated as single atoms with the approximate mass 16. There are indications that such a model for the immediate neighborhood of the carboxyl carbon should be able to present a useful vibrational pattern for the calculation of vibrational changes caused by changes in the mass of that carbon. Such a four-atom molecule will have six vibrational modes. Pitzer used force constants yielding approximately such frequency values with C^{12} as are observed with carboxylic acids. With the same force constants he also calculated the frequencies of the corresponding C^{14} molecule. Thus six terms $G(u) \Delta u$ were obtained.

In the transition state the bond R—C was assumed to be broken, and thus to have zero force constant. The two carbon-oxygen bonds were assumed to have their force constants unchanged from the reactant. All bending force constants were ascribed the value zero. Thus the pertinent part of the transition state was simply composed of the two carbon-oxygen oscillators, giving two $G(u^\ddagger) \Delta u^\ddagger$ terms. All other frequencies occurring in the molecule were assumed to disappear by cancellation between the reactant and the transition state.

Pitzer found the value 0.88 at 400°K for the relative rate of breaking $R—C^{14}$ and $R—C^{12}$ bonds, using the Slater reaction coordinate. The predicted isotope effect seems to be much too strong. Pitzer also discussed the experimental manifestations of a possible failure of the rate of activation to maintain equilibrium concentrations of all competing transition states. There seems, however, at present to be fairly little reason to doubt this very fundamental assumption.

Bigeleisen and Wolfsberg in their recent review (14) have discussed the general conclusions which could be drawn from the best experimental results available at present. The predicted figures for the ratio $2k_3''/k_1$ seem to be too high, and the experimental secondary isotope effect due to the presence of isotopic carbon in the intact carboxyl group seems to be quite real in spite of the fact that a combination of the fairly early results of Roe and Hellmann on one hand and Ropp and Raaen on the other seemed to indicate the contrary (see above). Loosening of the bond to this carbon atom on passing into the transition state should be able to account for such an effect. The treatment of Bigeleisen (9), already referred to, was of this kind but failed to do so, giving $2k_3''/k_1 = 0.997$ at 425°K, which seems rather peculiar. A much cruder treatment accounting only for the stretching frequencies of the carbon-carbon bonds gives a much more satisfactory figure. Picturing the model of the formation of the transition state in this case,

it is evident that only the stretching frequency of the isotopic bond in the reactant will give a contribution to the expression for the isotope effect. The transition states contain no bond to be accounted for, or, what amounts to the same, the stretchings have zero frequency, hence $u^{\ddagger} = 0$, and the other oscillator in the reactant is not isotopic, hence $\Delta u = 0$. The Slater coordinate yields identical reduced masses. Consequently we are left with only

$$\frac{2k_3''}{k_1} = 1 + G(u)\,\Delta u$$

where u and Δu as before correspond to the wave numbers 900 and -33 cm^{-1}, respectively. At 427°K this gives $2k_3''/k_1 = = 0.975$.

The problem of the decarboxylation of malonic acid has been treated at some length because it involves many problems the discussion of which might be useful as an illustration of the use of Bigeleisen's expression for heavy elements. It is also probably the problem which has hitherto been most thoroughly worked through. Related reactions have been investigated and have generally given results compatible with those for malonic acid.

It might seem somewhat disappointing that the crudest models have given predictions at least as good as those of more refined calculations. This is certainly not a fault of the general theory but emphasizes the very great difficulties inherent in a satisfactory vibrational treatment of molecules other than the very simplest ones, and especially in such a treatment of the transition state.

CARBON ISOTOPE EFFECT IN ALIPHATIC NUCLEOPHILIC SUBSTITUTION

The carbon isotope effects treated in previous sections pertain to reactions of pronounced decomposition type. The transition states could certainly be truly depicted as two-center aggregates. Besides three-center reactions on hydrogen, however, organic chemistry knows many examples of $S2$ reactions (Ingold's terminology), i.e., single-stage substitution reactions which are bimolecular. The transition state is made up of one residue, one leaving and one entering substituent. An outstanding family of such reactions are the S_N2 reactions on carbon, in which one entity with a lone electron pair replaces another at a carbon atom. The carbon atom is generally a member of an aliphatic molecule. The perfectly symmetrical reaction might be illustrated by halide ion exchange with an aliphatic halide:

$$(X^*:)^- + RX \; \rightleftarrows \; X^*R + (:X)^-$$

In order for such an exchange to be observable, isotopic halogen, here denoted by the presence or absence of an asterisk, has to be used. The reason why the carbon isotope effect of this reaction has not hitherto been investigated in spite of its very great interest is certainly that the problem is one requiring double-labeling technique, which would be difficult as well as expensive. In principle, pure samples of two isotopes of one of the isotopic elements involved, carbon or halogen, would provide the experimental possibilities.

Experimental Results. Working with two different nucleophiles, any mixture of two carbon isotopes is sufficient, and it is favorable from the kinetic point of view if one of them is present only in tracer amounts. Thus Bender and Hoeg (5) have investigated the reactions between methyl-C^{14} iodide and various nucleophiles, as hydroxide ion, triethylamine, and pyridine in solutions at 25°C. Since tracer amounts were used and the reaction is certainly of first order in the isotopic species, one of the simple equations, (3–4) or (3–7), could be used, which one depending upon whether the specific activity of the recovered reactant or the organic product was followed. In the same way Buist and Bender (23) determined the isotope effect in the reaction between methyl-C^{14} iodide and some aromatic amines of the N,N-dialkylaniline type at somewhat higher temperatures.

All these reactions are known to be of S_N2 mechanism, and the rates and activation energies cover a wide range. It was hoped that a possible correlation between isotope effect and activation energy would show up in the results. All values found were in the region $k_{14}/k_{12} = 0.876$ to 0.919, the temperatures being $25°$ to $63°C$. It is not probable that the range of isotope effects would be much larger if all observations had been carried out at the same temperature, because the weakest effect (with hydroxide ion) was observed at the same temperature, $25°C$, as the strongest one (with pyridine). There is some indication of a correlation of strong isotope effects with high energies of activation, but most of the differences fall within the limits of experimental uncertainty.

It was also hoped that the transition from S_N2 to S_N1 would be reflected by the strength of the isotope effect. For that reason Bender and Buist (4) investigated the slightly alkaline hydrolysis of 2-chloro-2-methylpropane-2-C^{14} in mixed dioxane and water at $25°C$. t-Butyl chloride is the simplest molecule known to react according to the S_N1 mechanism; this means that the rate-determining step is the formation of a carbonium ion that subsequently undergoes rapid reactions to form more stable entities, in the present case t-butyl alcohol and a minor amount of isobutene.

$$(CH_3)_3CCl \xrightarrow[\text{(slow)}]{} (CH_3)_3C^+ + Cl^-$$

$$(CH_3)_3C^+ \xrightarrow[\text{(fast)}]{} \text{products}$$

The transition state here is likely to be of the two-center type, and the medium is generally considered to act only via its electrostatic properties. S_N2 reactions could be imagined to pass over gradually into S_N1 reactions when the attacking reagent becomes nucleophilically weaker, the activation energy rises, and the halide molecule obtains a structure fit for making the corresponding carbonium ion stable enough. Of course, the dielectric properties of the solvent, too, play a very important role in the stabilization of the carbonium ion.

From the point of view of isotope effects, it could possibly be expected that the S_N1 reactions, being at one end of the transition series just described, would have the strongest isotope effects. The transition state is also likely to be fairly loose. The result with t-butyl chloride does not substantiate this, k_{14}/k_{12} being 0.974, but it should be remembered that neither the hydrocarbon part nor the leaving halogen is comparable to those with methyl iodide. Bender and Hoeg (5) also studied the reaction of methyl iodide with silver nitrate, a reaction considered to be a borderline case between S_N1 and S_N2. The observed isotope effect, $k_{14}/k_{12} = 0.921$ at 25°C, was nearly the same as with hydroxide ion.

Theoretical Predictions. The theoretical treatment of the isotope effect in these reactions offers several interesting problems. Bender and co-workers used the three-center relation of Bigeleisen and Wolfsberg (Eq. 2–22) for the calculation of $\nu_{L(1)}{}^{\ddagger}/\nu_{L(2)}{}^{\ddagger}$ in their different reactions. Y was identified with the organic fragment, Z with the halide ion, and X with the nucleophile in question. The parameter p was assigned the value 0 or 1, corresponding to a rate-determining dissociation of the halide (S_N1) or an approximately symmetrical three-center reaction (S_N2), respectively. The temperature-independent factor generally did not vary much with p in the interval $0 \le p \le 1$. Thus, for the reaction between methyl-C^{14} iodide and hydroxide ion, $m_{Y(1)} = 17$, $m_{Y(2)} = 15$, $m_{Z(1)} = m_{Z(2)} = 127$, $m_{X(1)} = m_{X(2)} = 17$,

$$\frac{\nu_{L(1)}{}^{\ddagger}}{\nu_{L(2)}{}^{\ddagger}} = 0.946 \quad \text{if } p = 0 \quad \text{and} \quad \frac{\nu_{L(1)}{}^{\ddagger}}{\nu_{L(2)}{}^{\ddagger}} = 0.952 \quad \text{if } p = 1$$

In the case of methyl iodide and pyridine we have to change the value of m_X to $m_{X(1)} = m_{X(2)} = 79$, and we obtain

$$\frac{\nu_{L(1)}{}^{\ddagger}}{\nu_{L(2)}{}^{\ddagger}} = 0.946 \quad \text{as before for } p = 0$$

and

$$\frac{\nu_{L(1)}{}^{\ddagger}}{\nu_{L(2)}{}^{\ddagger}} = 0.944 \quad \text{for } p = 1$$

If X, Y, and Z were instead identified with the pertinent atoms, the values would be 0.932, 0.939, 0.932, and 0.941, respectively. It is evident that the temperature-independent factor is not particularly sensitive to the assumptions about the model.

As to the temperature-dependent factor, it is necessary as usual to make extensive simplifications. Bender and Hoeg assume that all frequencies except the stretching frequency of the carbon-halogen bond to be opened are cancelled, either because they could be assumed to be equal in the transition state and the reactant, or, for those in the attacking nucleophile, they are non-isotopic. The frequency zero is ascribed to the carbon-halogen bond in the transition state. Thus what is left for the temperature-dependent factor of Eq. (2–19) is simply $1 + G(u_k) \, \Delta u_k$, where u_k refers to the stretching in the reactant.

There is one normal mode of vibration of methyl halides that could be described as involving mainly a vibration of the methyl group as a whole and the halogen against one another. The isotope shift on substitution of C^{13} for C^{12} in methyl iodide has been determined by Bernstein, Cleveland, and Voelz (7). Bender and Hoeg used the value 533.4 cm^{-1} for $C^{12}H_3I$, and 16.1 cm^{-1} for the corresponding shift for C^{13}/C^{12}. This shift agrees very well with that calculated from the reduced masses of CH_3 and I, and the movement is certainly closely related to a separation of the two groups as in the present reaction. In order to obtain the temperature-dependent factor for C^{14}/C^{12}, it is possible simply to double the shift. Thus the factor is found to be 0.970 at 25°C and 0.976 at the highest temperature used, 63°C.

If the value 0.970 for the temperature-dependent factor is combined with the lowest value of the temperature-independent factor, 0.932, we obtain $k_{14}/k_{12} = 0.904$, and, with the highest value of the last-mentioned factor, 0.952, $k_{14}/k_{12} = 0.923$. The observed isotope effects are of about this strength or somewhat stronger.

It could be questioned, however, whether Bigeleisen's formalism is the best approach in this particular case. It might be of interest to use, for instance, Eq. (2–6), which also rests on a single-frequency approximation. The difference is seen, from Eqs. (2–14) and (2–15), to consist in calculating the factors containing molecular masses and moments of inertia directly or replacing them by a product of frequency ratios that is simplified later on. The particular geometric shape of the aggregates in the present case makes them fit for a direct calculation of the former dynamical quantities. The remaining approximations are the same according to both methods, and, owing to the fact that no frequencies within the large products of Eq. (2–5) are particularly low, the vibrational partition function ratios of the type $(1 - e^{-u(2)})/(1 - e^{-u(1)})$ could be expected to cause very small deviations from unity even if cancellation is not entirely perfect.

The transition states in the S_N2 reactions are certainly of the type

$$
\begin{array}{c}
\text{H} \qquad \text{H} \\
\diagdown \quad \diagup \\
\text{HO} ---- \text{C} ---- \text{I} \\
\mid \\
\text{H}
\end{array}
$$

where oxygen, carbon, and iodine are situated on a straight line and the methyl group is approximately planar. It seems that a good approximation would be to treat the body as a linear three-atom molecule,

$$ \text{X} ---- \text{Y} ---- \text{Z} $$

in this particular case with the atomic masses 17, 15 (17), and 127, respectively. The real molecule has one moment of inertia which is non-isotopic and corresponds to the vanishing one in the linear three-atom model. The other two moments of inertia are equal and isotopic. Only the latter need be included in the calculations and might be written

$$ I^\ddagger = $$

$$ = \frac{1}{M^\ddagger}\left[m_X(m_Y + m_Z)r_{XY}^2 + 2m_Xm_Zr_{XY}r_{YZ} + m_Z(m_X + m_Y)r_{YZ}^2\right] $$

where $M^\ddagger$ is the mass of the transition state, equal to $(m_X + m_Y + m_Z)$, and r denotes interatomic distance.

The moments of inertia of the reactant in the same way consist of one non-isotopic and two equal and isotopic ones.

In the diatomic molecule approximation the latter can be written

$$I = \frac{m_Y m_Z}{M} (r_{YZ}^0)^2$$

where M denotes the molecular mass $(m_Y + m_Z)$ and r_{YZ}^0 is the distance between the atoms in the unperturbed molecule.

With these quantities the two parentheses (or the temperature-independent factor) of Eq. (2–6) become

$$\left(\frac{M_1^{\ddagger}}{M_2^{\ddagger}} \times \frac{M_2}{M_1}\right)^{3/2} \times \frac{I_1^{\ddagger}}{I_2^{\ddagger}} \times \frac{I_2}{I_1} = \left(\frac{M_1^{\ddagger}}{M_2^{\ddagger}} \times \frac{M_2}{M_1}\right)^{1/2} \times$$

$$\times \frac{m_X(m_{Y(1)} + m_Z)r_{XY}^2 + 2m_X m_Z r_{XY} r_{YZ} + m_Z(m_X + m_{Y(1)})r_{YZ}^2}{m_X(m_{Y(2)} + m_Z)r_{XY}^2 + 2m_X m_Z r_{XY} r_{YZ} + m_Z(m_X + m_{Y(2)})r_{YZ}^2} \times$$

$$\times \frac{m_{Y(2)}}{m_{Y(1)}}$$

because only the atom Y is isotopic.

The problem now is to choose reasonable values for the distances r_{XY} and r_{YZ} in the transition state. The oxygen-carbon distance in methanol is of the order of 1.4 Å, and the carbon-iodine distance in methyl iodide is of the order of 2.1 Å (100). Only the ratio r_{XY}/r_{YZ} need be estimated, and it seems safe to try on one hand $r_{XY}/r_{YZ} = 1$ and on the other $r_{XY}/r_{YZ} = 0.5$. Denoting by the index 1 the case with C^{14} and by 2 the case with C^{12}, the above expression becomes 0.905 and 0.916, respectively.

The last term of Eq. (2–6) is

$$\frac{\sinh \frac{1}{2} u_{k(1)}}{\sinh \frac{1}{2} u_{k(2)}}$$

where u_k corresponds to the single vibration which is not assumed to be cancelled. With the frequencies 533.4 and $(533.4 - 2 \times 16.1) = 501.2$ cm^{-1}, this gives 0.912 at 25°C.

The two factors give $k_{14}/k_{12} = 0.825$ or 0.835, dependent on the choice of the first one. This isotope effect is more than twice as strong as that obtained from Bigeleisen's and Bigeleisen-Wolfsberg's expressions. For the same model the latter was $k_{14}/k_{12} = 0.923$. Since the observed value for the hydroxide reaction was 0.919 and the value calculated from Eq. (2–6) is likely to be something of an upper limit of the strength, it seems the agreement is not too bad.

If the complete set of normal vibrations were used, the two treatments should certainly give the same prediction. Neither the Teller-Redlich product rule nor the approximations in Bigeleisen's treatment of the fundamental equation should be able to cause such serious discrepancies. The reason is the rough approximation in the replacement of the molecular masses and the moments of inertia by an oversimplified product of frequency ratios. If Eq. (2–15) is divided by Eq. (2–14) we obtain

$$\left(\frac{M_1^\ddagger}{M_2^\ddagger} \times \frac{M_2}{M_1}\right)^{3/2} \left(\frac{A_1^\ddagger B_1^\ddagger C_1^\ddagger}{A_2^\ddagger B_2^\ddagger C_2^\ddagger} \times \frac{A_2 B_2 C_2}{A_1 B_1 C_1}\right)^{1/2} =$$

$$= \frac{\nu_{L(1)}^\ddagger}{\nu_{L(2)}^\ddagger} \prod_i^{3n^\ddagger-7} \frac{\nu_{i(1)}^\ddagger}{\nu_{i(2)}^\ddagger} \prod_i^{3n-6} \frac{\nu_{i(2)}}{\nu_{i(1)}}$$

since the atomic masses disappear by cancellation as explained on p. 34.

In our particular example, the left-hand side of the expression above was found to give a value somewhere between 0.905 and 0.916 when index 1 corresponded to C^{14}, and 2 to C^{12}. According to the assumptions concerning the vibrational frequencies, most of the right-hand side is cancelled out, leaving only

$$\frac{\nu_{L(1)}^\ddagger}{\nu_{L(2)}^\ddagger} \times \frac{\nu_{k(2)}}{\nu_{k(1)}} = 0.952 \times \frac{533.4}{501.2} = 1.013$$

The approximate treatment of the frequency ratios thus tends to increase the ratio k_{14}/k_{12} by a factor of about 1.11. This is also the ratio between the two predictions above.

The transition state of S_N2 reactions offers an exceptionally favorable opportunity for a direct calculation of the moments of inertia when the organic residue is as simple as methyl. In general, the conformation of the transition state is very little known, and, even when it is known in principle, the computation of the moments of inertia is often much more difficult and uncertain. In the present case the uncertainty is of the order of 1 per cent. If it had been ten times larger, it would have been of the same order as the uncertainty in the frequency expression. In most practical cases it is certainly larger than that, and then the ordinary use of Bigeleisen's expression is probably the best approach. In the decarboxylation of malonic acid, for instance, it would be fairly hopeless to try to compute the principal moments of inertia. The case we have treated here, however, may serve as a warning that the accuracy of such predictions should not be overestimated. It makes it easy also to see how manipulation with the frequencies, like the one by Pitzer referred to in the discussion of the decarboxylation of malonic acid (p. 141), might cause considerable changes in the predictions.

The terms "temperature-dependent" and "temperature-independent" are frequently used for different contributions to the ratio k_1/k_2. It should be observed that there is no unambiguous correspondence between the respective factors in the two treatments, because according to Bigeleisen's method the frequency ratios are included in the "temperature-dependent" factor for the sake of convenience, although these ratios are in fact independent of the temperature.

Owing to the slight sensitivity of the Bigeleisen-Wolfsberg temperature-independent factor to changes of the parameter p between 0 and 1, that theory gave very little hope of finding a difference in isotope effect between S_N1 and S_N2 reactions. Depending on the relative masses of the leaving and the entering group, either type of reaction could give the strongest effect, and the interval of variation was narrow. The Bigeleisen temperature-dependent factor was the same for either mechanism.

The treatment involving the actual moments of inertia furnishes a different prediction. If the S_N1 reaction is pictured as a true unimolecular decomposition of the diatomic molecule YZ, the molecular masses and the moments of inertia will be cancelled from Eq. (2–5) or Eq. (2–6). The molecular masses are the same in the reactant and the transition state of each isotopic case. The moment of inertia of a diatomic molecule is the product of the reduced mass and the square of the inter-atomic distance. The reduced masses are the same in the reactant and the transition state of each isotopic case and hence are cancelled in the same way as the molecular masses. The squared distance is the same for the two isotopic transition states and thus disappears by cancellation. In the same way the squared distance disappears for the reactants. (The cancellation can also be seen if m_X is assigned the value 0 in the expression on p. 149.) We are left with the temperature-dependent factor $(\sinh \frac{1}{2} u_{k(1)})/(\sinh \frac{1}{2} u_{k(2)})$, which has been found to give $k_{14}/k_{12} = 0.912$.

Taking the actual moments of inertia, although approximate, into consideration, we thus obtain the prediction of an isotope effect about half as strong for the S_N1 as for the S_N2 mechanism. This agrees qualitatively with the results of Bender and Buist (4), who obtained $k_{14}/k_{12} = 0.974$ at 25°C for the hydrolysis of t-butyl chloride, if the comparison of this compound with methyl iodide were not somewhat doubtful. Anyhow, in the light of this result and the last method of computation, the hope that the carbon isotope effect might serve as a diagnostic tool in the discrimination between S_N1 and S_N2 reactions seems strengthened, although it should be admitted that the theoretical difference between S_N1 and S_N2 mechanisms, caused mainly by the ratio between the moments of inertia of the isotopic transition states, becomes less with increasing complexity of the latter. The problem is certainly worth further investigation, and such studies could conveniently be combined with measurements of the α-hydrogen secondary isotope effect (cf. pp. 90 ff.).

It can easily be seen that Bigeleisen's treatment with the usual simple approximations will be a much better approach to unimolecular decompositions than to the S_N2 reaction just discussed. From what has just been said it follows that the molecular masses and the moments of inertia give the ratio unity according to the diatomic model. Hence according to the relation on p. 150 the complete product of frequency ratios should also be equal to unity. Generally $\nu_{L(1)}{}^{\ddagger}/\nu_{L(2)}{}^{\ddagger}$ will not differ much from $\nu_{k(1)}/\nu_{k(2)}$. Thus, putting the product of the remaining ratios equal to unity, as is done in the computation of the isotope effect, is certainly a good approximation in this case.

It has been pointed out that solvation might introduce considerable uncertainty into the present calculations and particularly in the comparison between S_N1 and S_N2 reactions, for which types of reaction its role is likely to be different. Owing to the equivalence of Eqs. (2–5) and (2–16), this criticism strikes equally the two computational methods compared above, neither of which refers to the solvated species. If the solvation sphere were to be included in the calculations, the molecular masses and moments of inertia would have to be changed in one type of computation, and this would correspond to the introduction of additional frequency ratios in the other. Moreover, the number of factors containing exponential expressions would have to be increased identically in both computations. Obviously our present information is quite insufficient for such a complete treatment.

Since all computations of specific rate ratios carried out above refer to models without solvation, the best we could do, if it were possible, would be to multiply the above figures by correction factors accounting for the deviation between perfect computations of the rate ratios based on solvated models and such computations based on models without solvation. Since perfect calculations are independent of whether Eq. (2–5) or Eq. (2–16) is used, we would obtain one single factor for the S_N1 and one for the S_N2 reaction. It is obvious from the values obtained above that the two rate ratios computed from the molecular masses and the moments of inertia are more likely than the others to give a difference between S_N1 and S_N2 reactions in the observed direction even after correction, although it cannot be completely excluded that the correction factors might be so different that the strengths of the isotope effects will differ in the wrong direction for both sets of values.

It is difficult to assess the error introduced by neglecting solvation. If the mutual mobility between the chemical species in question and their shells of solvating molecules is not too limited, it might be fairly unimportant. In the present case it is fortunate that the S_N2 transition state, which is an ion and hence could be expected to be subject to particularly strong solvation, has moments of inertia which are only slightly sensitive to the isotopic mass of the carbon already without solvation.

8

Carbon Isotope Effects and Reaction Mechanisms

As seen in the foregoing chapter, the presence of isotopic carbon at a bond which is ruptured in a reaction generally causes a measurable isotope effect. Certainly the measurements and the purifications of samples require a good deal of care, but undoubtedly the carbon isotope effect should be able to serve as a tool in unraveling complicated reaction mechanisms, mainly in the same way as the hydrogen isotope effect. Carbon isotope effects are small, but, on the other hand, compounds isotopic in carbon should also be less susceptible to isotope effects of a more physical kind than compounds isotopic in hydrogen. For instance, compounds likely to form hydrogen bonds could be expected to show isotope effects in distillation and crystallization when the pertinent hydrogen is isotopic. Generally such hydrogen is also easily exchangeable, and this phenomenon tends to render isotope effect investigations impossible, but there are probably cases where isotope effects of more physical origin introduce uncertainties, particularly in reactions which show no inherent isotope effect. Isotopic carbon atoms have less relative mass differences, and carbon atoms are much more protected in ordinary molecules than are hydrogen atoms. Specific interactions between the carbon atoms of an ordinary stable molecule and the surrounding medium, similar to hydrogen bonding, are hardly conceivable.

155

It is therefore probable that the products of reactions having no inherent isotope effect will also show no isotope discrimination up to a high degree of accuracy.

ACID DECARBONYLATION OF CARBOXYLIC ACIDS

An example of the way carbon isotope effect studies may serve as a tool in the elucidation of complex reaction mechanisms is a recent discussion by Ropp (81) of the acid decarbonylation of carboxylic acids. The problem has already been touched upon in the discussion of the decarbonylation of formic acid (pp. 128 ff.).

The correlation between reaction rate and medium composition for the decarbonylation of triphenylacetic acid in sulfuric acid led Deno and Taft (29) to propose the following mechanism:

$$Ph_3CCOOH + H^+ \rightleftarrows Ph_3CCO^+ + H_2O \quad \text{(equilibrium)}$$
$$Ph_3CCO^+ \rightarrow Ph_3C^+ + CO \quad \quad \text{(rate-determining)}$$

Ropp has found two pieces of isotopic evidence which contradict this scheme. Heavy oxygen equilibrates incompletely between the sulfuric acid and the carboxylic acid, the evolved carbon monoxide having an O^{18} content of only about one-fifth of the equilibrium one. The claim for complete equilibration arises from the existence of a symmetrical form of the protonated acid, $RC(OH)_2^+$. The other piece of evidence is the absence of a measurable isotope effect in the decarbonylation of triphenyl-acetic-2-C^{14} acid, indicating that the rupture of the carbon-carbon bond can hardly be entirely rate-determining.

Ropp suggests the following general reaction scheme for the decarbonylation of acids:

(A) $RCOOH + H_2SO_4 \rightleftarrows RC{\overset{\displaystyle O}{\underset{\textstyle OH_2^+}{<}}} + HSO_4^-$ (equilibrium)

(B) $RCOOH_2^+ \rightarrow RCO^+ + H_2O$ (at least partly rate-determining)

(C) $RCO^+ + H_2O \rightarrow RCOOH_2^+$

(D) $RCO^+ + H_2SO_4 \rightarrow (R \cdot H_2SO_4)^+ + CO$
                (C and D competing with one another)
(E) $(R \cdot H_2SO_4)^+ + 2H_2O \rightarrow ROH + H_3O^+ + H_2SO_4$

In the case of triphenylacetic acid the incomplete oxygen exchange is likely to be due to some competition of reaction (C), which is the reversal of (B), with (D), which could be expected to be comparatively slow owing to steric hindrance in the attack of sulfuric acid on the central carbon of the triphenylmethyl group. Reaction (C) cannot, however, be more rapid than (D), because the oxygen exchange was far from complete. The observed correlation between rate and medium composition might easily be explained by neither (B) nor (D) being completely rate-determining.

With formic acid, reaction (D) implies a simple hydrogen transfer and is certainly much faster than (C). This agrees with the fact that the oxygen of the carbon monoxide is not at all equilibrated with the oxygen of the sulfuric acid. The large carbon isotope effect already discussed (pp. 128 ff.) arises, of course, in step (B). Also, the dependence of the rate on the medium composition is compatible with the scheme. Finally, an observed secondary deuterium isotope effect ($k_D/k_H = 0.67$) is probably due to a weakening of the carbon-hydrogen bond in the transition state of reaction (B).

9

Isotope Effects with Elements Heavier than Carbon

In Chapter 7 we saw that a relative mass difference as small as $1:12$ is able to give rise to measurable isotope effects. With increase in the accuracy of the quantitative determinations of isotope ratios in chemical substances, it ought to be possible to study much heavier elements, because all secondary effects tending to introduce false isotope fractionation, for instance through necessary preparative handling of samples, will certainly decrease simultaneously with the main effect.

In the present chapter examples of studies of isotope effects at oxygen and sulfur will be discussed.

OXYGEN ISOTOPE EFFECT IN THE THERMAL DECOMPOSITION OF AMMONIUM NITRATE

Friedman and Bigeleisen (36) have studied the isotopic behavior of nitrogen and oxygen in the thermal decomposition of ammonium nitrate. As a consequence of several pieces of experimental evidence, which will not be reviewed here, they propose a two-step dehydration mechanism for the reaction:

$$NH_4^+ + NO_3^- \longrightarrow H_2N—NO_2 + H_2O$$
$$H_2N—NO_2 \longrightarrow N_2O + H_2O$$

Owing to the more complicated role of nitrogen, no effort has been made to calculate its isotope effect. Oxygen, on the other hand, is taken away from nitrogen in both steps, and the isotope effect might be considered roughly the same in both of them. The results allow some interesting conclusions to be drawn about the vibrational behavior of the transition state of the first reaction step, i.e., a nitrate ion which is losing one of its oxygens.

Friedman and Bigeleisen decomposed ammonium nitrate of ordinary isotope composition and studied the initial isotope composition of the nitrous oxide evolved and the final over-all isotope composition of both products. The reaction temperature was $220° \pm 20°C$.

The following reaction scheme forms the basis of the computations (no mass number on oxygen indicates the ordinary one, 16):

$$NH_4^+ + NO_3^- \xrightarrow{k_1} H_2NNO_2 + H_2O$$

$$NH_4^+ + (NO_2O^{18})^- \begin{cases} \xrightarrow{k_1'} H_2NNO_2 + H_2O^{18} \\ \xrightarrow{k_1''} H_2NNOO^{18} + H_2O \end{cases}$$

$$H_2NNO_2 \xrightarrow{k_2} N_2O + H_2O$$

$$H_2NNOO^{18} \begin{cases} \xrightarrow{k_2'} N_2O + H_2O^{18} \\ \xrightarrow{k_2''} N_2O^{18} + H_2O \end{cases}$$

(Owing to the small natural abundance of O^{18}, molecules containing more than one heavy oxygen can be neglected.)

If the first reaction step is assumed to be rate-determining, there will be no piling up of the intermediate. If the initial amounts of $NH_4NO_2O^{18}$ and ordinary NH_4NO_3 are denoted by a_1^0 and a_2^0, respectively, it is easy to see that

$$\frac{\left(\dfrac{\text{amount of } N_2O^{18}}{\text{amount of } N_2O}\right)_{\text{initial}}}{\left(\dfrac{\text{amount of } N_2O^{18}}{\text{amount of } N_2O}\right)_{\text{total, final}}} = \frac{\dfrac{\dfrac{k_2''}{k_2' + k_2''}k_1''a_1^0}{k_1 a_2^0}}{\dfrac{\dfrac{k_2''}{k_2' + k_2''} \times \dfrac{k_1''}{k_1' + k_1''}a_1^0}{a_2^0}} = \frac{k_1' + k_1''}{k_1}$$

where the amount of light compound generated by the heavy reactant has been neglected at the side of the bulk generated from the light one. It is interesting that the ratio computed above gives information about the first reaction step only.

Another value measured was

$$\frac{\left(\dfrac{\text{amount of } N_2O^{18}}{\text{amount of } N_2O}\right)_{\text{total, final}}}{\left(\dfrac{\text{amount of } H_2O^{18}}{\text{amount of } H_2O}\right)_{\text{total, final}}} =$$

$$= \frac{\dfrac{\dfrac{k_2''}{k_2' + k_2''} \times \dfrac{k_1''}{k_1' + k_1''} a_1^0}{a_2^0}}{\dfrac{\dfrac{k_1'}{k_1' + k_1''} a_1^0 + \dfrac{k_2'}{k_2' + k_2''} \times \dfrac{k_1''}{k_1' + k_1''} a_1^0}{2a_2^0}} =$$

$$= \frac{2 \dfrac{k_2''}{k_2' + k_2''} \times \dfrac{k_1''}{k_1' + k_1''}}{\dfrac{k_1'}{k_1' + k_1''} + \dfrac{k_2'}{k_2' + k_2''} \times \dfrac{k_1''}{k_1' + k_1''}}$$

where the same approximation as above has been used.

The reactants and the transition states in the three possible modifications of the first step may be represented schematically in the following way:

(spec. rate = k_1)   (spec. rate = k_1')

(spec. rate = k_1'')

Giving due attention to symmetry numbers and denoting the frequencies in the same way as the specific rates, we obtain, according to Eq. (2–19),

$$\frac{k_1'}{k_1} \times \frac{6}{2} \times \frac{2}{2} =$$

$$= \frac{\nu_L^{\ddagger'}}{\nu_L^{\ddagger}} \left[1 - \Sigma G(u_1{}^{\ddagger})(u_1{}^{\ddagger'} - u_1{}^{\ddagger}) + \Sigma G(u_1)(u_1' - u_1) \right]$$

$$\frac{k_1''}{k_1} \times \frac{6}{2} \times \frac{1}{2} =$$

$$= \frac{\nu_L^{\ddagger''}}{\nu_L^{\ddagger}} \left[1 - \Sigma G(u_1{}^{\ddagger})(u_1{}^{\ddagger''} - u_1{}^{\ddagger}) + \Sigma G(u_1)(u_1'' - u_1) \right]$$

where the sums are taken over all real vibrational modes. Of course, $u'' = u'$ for the same reactant, but for the sake of symmetry we keep both symbols. From the above expressions there follows

$$\frac{k_1' + k_1''}{k_1} =$$

$$= \frac{1}{3} \times \frac{\nu_L^{\ddagger'}}{\nu_L^{\ddagger}} \left[1 - \Sigma G(u_1{}^{\ddagger})(u_1{}^{\ddagger'} - u_1{}^{\ddagger}) + \Sigma G(u_1)(u_1' - u_1) \right] +$$

$$+ \frac{2}{3} \times \frac{\nu_L^{\ddagger''}}{\nu_L^{\ddagger}} \left[1 - \Sigma G(u_1{}^{\ddagger})(u_1{}^{\ddagger''} - u_1{}^{\ddagger}) + \Sigma G(u_1)(u_1'' - u_1) \right]$$

Experimentally, the isotopic composition of the nitrous oxide was found to be the same initially and finally within the accuracy of the measurements. Thus, according to the relation derived above, $(k_1' + k_1'')/k_1 = 1$. If this value is introduced into our last expression, the following empirical equation is obtained:

$$\frac{1}{3} \times \frac{\nu_L^{\ddagger'}}{\nu_L^{\ddagger}} \left[\Sigma G(u_1{}^{\ddagger})(u_1{}^{\ddagger} - u_1{}^{\ddagger'}) - \Sigma G(u_1)(u_1 - u_1') \right] +$$

$$+ \frac{2}{3} \times \frac{\nu_L^{\ddagger''}}{\nu_L^{\ddagger}} \left[\Sigma G(u_1{}^{\ddagger})(u_1{}^{\ddagger} - u_1{}^{\ddagger''}) - \Sigma G(u_1)(u_1 - u_1'') \right] =$$

$$= 1 - \frac{1}{3} \times \frac{\nu_L^{\ddagger'}}{\nu_L^{\ddagger}} - \frac{2}{3} \times \frac{\nu_L^{\ddagger''}}{\nu_L^{\ddagger}}$$

Evaluation of the two frequency ratios according to Eq. (2–20) gives $\nu_L^{\ddagger\prime}/\nu_L^{\ddagger} = 0.974$, $\nu_L^{\ddagger\prime\prime}/\nu_L^{\ddagger} = 1$, and, if the leaving oxygen is assumed to be loaded with two hydrogens, the figures become 0.978 and 1, respectively. The right-hand side of the empirical equation becomes in this way 0.009 or 0.007. In view of the uncertainty in the experimentally determined figure 1, the appropriate limit of accuracy seems to be ±0.005, making the value of the right-hand side somewhere between $+0.002$ and $+0.014$. The left-hand side of the empirical equation has thus been found to be slightly positive.

The left-hand side could be considered a kind of weighted mean for the difference in the sum $\sum G(u)\,\Delta u$ between the transition state and the reactant. The sum of the weights is not exactly unity, but this is of little importance in the present estimations. The relation does not mean, of course, that each term in the transition state sum is always larger than a corresponding term in the reactant sum, nor even that both square brackets above need to be positive. It should be remembered that one degree of vibrational movement corresponds to motion along the decomposition coordinate and is excluded from the sum of the transition state. According to the very crude picture with pure "bond oscillators," the stretching of one of the nitrogen-oxygen bonds in the reactant has no counterpart in the transition state sum. For the case of the first square bracket this particular bond is isotopic, giving a large positive $u_1 - u_1'$. It would not be astonishing to find the expression within this bracket negative.

As to the second square bracket, the vibrations of the isotopic bonds occur in both sums, and the broken bond is not isotopic. That there are different numbers of terms in the two sums is thus no reason why the expression should be algebraically small, because the extra term has $u_1 - u_1'' \approx 0$. In order for this bracket to be positive and capable of more than outweighing the probably negative first one, the transition state sum must contain some term or terms for which $G(u)\,\Delta u$ is appreciably larger than the corresponding terms of the reactant sum. This is most naturally explained by the assumption that those transition state vibrations, which are not to be excluded, have on the average higher frequencies than the vibrations of the reactant.

The last-mentioned conclusion might be expressed in another way. Experimentally we have found that $k_1' + k_1'' = k_1$, which means that nitrate ions containing O^{18} react with the same total reaction rate as ordinary nitrate ions. If $k_1' < k_1/3$, which seems likely, k_1'' must be so much larger than $2k_1/3$. This is a result of a favorable change in equilibrium position for the transition state relative to the reactant, brought about by the introduction of the heavy isotope. It is well known that heavy isotopes are enriched in that member of an exchange equilibrium which has the highest vibrational frequencies. Thus it seems probable that large frequencies of the transition state tend to enrich the heavy isotope in it relative to the reactant, in this manner tending to speed up the average reaction rate of the heavy compound to the value of the rate of the ordinary one, although one mode of decomposition of the heavy reactant certainly operates in the opposite direction.

Friedman and Bigeleisen make an estimate of the equilibrium constant of the isotope exchange equilibrium:

(heavy reactant) + (light trans.st.) $\rightleftarrows$
$$\rightleftarrows \text{(light reactant)} + \text{(heavy trans.st.)}$$

by means of the above-mentioned figures. According to pp. 34–35 and a comparison of Eqs. (2–16) and (2–19), it is obvious that this equilibrium constant could be written

$$1 - \sum G(u_l^\ddagger)(u_h^\ddagger - u_l^\ddagger) + \sum G(u_l)(u_h - u_l) =$$
$$= 1 + \sum G(u_l^\ddagger)(u_l^\ddagger - u_h^\ddagger) - \sum G(u_l)(u_l - u_h)$$

where indices h and l denote "heavy" and "light."

Our transition state consists of two forms which are isotopic isomers among themselves. Since the isotopic oxygen does not occupy equivalent positions, a weighted mean has to be taken in the calculation of the (average) equilibrium constant. The left-hand side of our empirical equation on p. 161 might serve as such a mean for the two sums of the equilibrium constant.

From the numerical value found above it is evident that the equilibrium constant must be within the interval 1.002 to 1.014.

Friedman and Bigeleisen point out that this value should be comparable to the corresponding one for the exchange of one heavy oxygen atom between the two stable entities carbonate ion and carbon dioxide,

$$\tfrac{1}{3}(CO_3^{18})^{2-} + \tfrac{1}{2}CO_2 \rightleftarrows \tfrac{1}{3}CO_3^{2-} + \tfrac{1}{2}CO_2^{18}$$

Urey (96) has calculated the equilibrium constant of the latter exchange reaction from vibrational data and found it to be 1.013 at 500°K. In comparison with this value the above figure seems entirely reasonable, particularly as our transition state has one degree of vibrational freedom which behaves classically and thus works in the direction opposite to heavy isotope enrichment.

In order to compute the other measurable ratio, intramolecular isotope effects are required. For instance, in the case of the ratio $2k_1'/k_1''$ the temperature-independent factor $\nu_L^{\ddagger'}/\nu_L^{\ddagger''}$ will have the above values, 0.974 or 0.978. The simplest possible calculation of the temperature-dependent factor according to Bigeleisen, taking only the stretching of the isotopic bond into consideration, gives the value 0.982 at 500°K. Together the two factors give $2k_1'/k_1'' = 0.956$ or 0.960.

If the mechanism of the second step is analogous to that of the first step and the two oxygen atoms of the intermediate occupy equivalent positions, $k_2'/k_2'' = 2k_1'/k_1''$. If we call this ratio α, the measured ratio between the final isotope content ratios of nitrous oxide and water should be equal to $4/\alpha(\alpha + 3)$. If the value $\alpha = 0.96$ is introduced, we obtain the value 1.052. With $\alpha = 0.98$ as a lower limit of the strength of the isotope effect, 1.026 is obtained. The experimental value was 1.023 ± 0.003.

SULFUR ISOTOPE EFFECT IN THE REDUCTION OF SULFATE ION

A study of a kinetic isotope effect at sulfur (S^{34}/S^{32}) has been carried out by Harrison and Thode (47). Sulfate ion of natural isotopic composition was reduced to hydrogen sulfide with a mixture of hydriodic, hypophosphorous, and hydrochloric acid at the temperatures 18°, 35°, and 50°C. The reaction is considered to consist of a reduction of the sulfate by iodide ions. The hypophosphorous acid serves to reduce the produced iodine to iodide again. Experiments with sulfite ion showed that the reduction of sulfate to sulfite ion is probably the rate-controlling step. Provided the subsequent reduction steps are quantitative, the isotopic composition of the hydrogen sulfide is representative of the isotope fractionation in the first step.

Comparisons of the isotopic composition of the hydrogen sulfide evolved during the first 2 per cent conversion with that of the sulfate gave directly the ratio $k_{34}/k_{32} = 0.978$, which was independent of the temperature and the concentration of hydriodic acid within the limits of experimental accuracy.

A computation of the temperature-independent factor according to Slater gives a value of 0.990. For the computation of a limit of the temperature-dependent factor in Bigeleisen's expression, Harrison and Thode used the sulfite ion as a model of the transition state. Taking all vibrational frequencies of the latter and all of the sulfate ion into account, they obtained for the total isotope effect $k_{34}/k_{32} = 0.963$ at 0°C, and 0.966 at 25°C. The agreement with the experimental value is good.

Since the sulfate ion is tetrahedral and the isotopic atom is the central one, it might be possible to treat it according to a simplified method proposed by Bigeleisen and Goeppert Mayer (12) for the computation of isotope exchange equilibrium constants. If there is an isotope exchange equilibrium,

$$A^*X_p + AX_q \rightleftarrows AX_p + A^*X_q$$

where A and A* are light and heavy isotopes, respectively, of a heavy element, and the ligand X surrounds A symmetrically in both kinds of molecules, it is possible to obtain an approximate value of the equilibrium constant from the expression

$$K \approx 1 + \frac{m_X \, \Delta m_A}{24 m_A^2} (q u_q^2 - p u_p^2)$$

where m_X is the mass of the ligand, m_A that of A, and $(m_A + \Delta m_A)$ that of A*. u_q and u_p are hc/kT times the symmetrical "breathing" frequencies (in wave number) of the molecules AX_q and AX_p, respectively. The u's may not be too large for this approximation to hold.

The sulfite ion has only three ligands surrounding sulfur and is pyramidal and not suitable for this kind of treatment, but it might still be worth trying the above formula with the number of ligands reduced to three in the transition state and assuming that the force constants for the bonds to these are unchanged and correspond to the frequency 981 cm^{-1} (48). The exchange equilibrium between the reactant and the transition state

$$S^{34}O_4^{2-} + (SO_3)_\ddagger \rightleftarrows SO_4^{2-} + (S^{34}O_3)_\ddagger$$

will on these assumptions have the equilibrium constant

$$K = 1 + \frac{16 \times 2}{24 \times (32)^2} (3 - 4)\left(\frac{hc \times 981}{k \times 300}\right)^2 = 0.971$$

at 300°K. According to the discussion on pp. 34–35, this equilibrium constant has to be multiplied by the ratio of the decomposition frequencies in order to give the rate-constant ratio. Thus

$$\frac{k_{34}}{k_{32}} = \frac{\nu_{L(34)}^\ddagger}{\nu_{L(32)}^\ddagger} K = 0.990 \times 0.971 = 0.961$$

at 300°K. This agrees very well with the prediction of Harrison and Thode, although it should be observed that the approximations become fairly rough with u values as high as the present one.

Owing to the high symmetry of the reactant, which is tetrahedral with the isotopic atom in the center of gravity, it could be of interest to apply Eq. (2–7). If the reaction is treated as a unimolecular decomposition, the molecular masses will cancel each other in the more elaborate Eq. (2–6). The lengthening of one of the sulfur-oxygen bonds will hardly displace the sulfur appreciably from the center of gravity in the transition state, and the moments of inertia will cancel each other almost exactly.

The possible presence of one or two hydrogen atoms at the pertinent oxygen in the transition state does not change these conclusions significantly.   The remainder of Eq. (2–6) is simply Eq. (2–7), the approximation of Eyring and Cagle:

$$\frac{k_{34}}{k_{32}} = \frac{\sinh \frac{1}{2} u_{k\,(34)}}{\sinh \frac{1}{2} u_{k\,(32)}}$$

The "breathing" vibration of the sulfate ion has the frequency 981 cm^{-1} (48).   This vibration involves the stretching of sulfur-oxygen bonds.   For the heavy molecule we calculate the corresponding frequency simply from the reduced masses of the isotopic diatomic S—O molecules, obtaining 971 cm^{-1}.   (It might seem strange to use different frequencies for different masses of the sulfur atom in view of the fact that the symmetric "breathing" vibration is independent of the mass of the central atom.   It is, however, a way of accounting for all the frequencies which are really sensitive to that mass but have been assumed to be cancelled.)   From these figures $k_{34}/k_{32} = 0.976$ at 300°K, in very good agreement with the experimental result.   The computation is, of course, as usual entirely equivalent to a Bigeleisen calculation on the same assumptions regarding the frequencies.

References

1. Bassett, I. M., and R. D. Brown. "A Theoretical Investigation of the Chemical Reactivity of Glyoxaline," *J. Chem. Soc.*, **1954**, 2701.
2. Bell, R. P. *Acid-Base Catalysis*, Clarendon Press, London, 1941, p. 143.
3. Bellamy, L. J. *The Infra-red Spectra of Complex Molecules*, 2d ed., Methuen & Co., Ltd., London; John Wiley & Sons, Inc., New York, 1958, (a) p. 13, (b) p. 34, (c) p. 75.
4. Bender, M. L., and G. J. Buist. "Carbon-14 Kinetic Isotope Effects. III. The Hydrolysis of 2-chloro-2-methylpropane-2-C^{14}," *J. Am. Chem. Soc.*, **80**, 4304 (1958).
5. Bender, M. L., and D. F. Hoeg. "Carbon-14 Kinetic Isotope Effects in Nucleophilic Substitution Reactions," *J. Am. Chem. Soc.*, **79**, 5649 (1957).
6. Berglund-Larsson, U., and L. Melander. "Isotope Effect of Hydrogen and Mechanism of Aromatic Sulphonation," *Arkiv Kemi*, **6**, 219 (1953); U. Berglund-Larsson. "Isotope Effect of Hydrogen in Aromatic Sulphonation. II. Temperature Dependence," *ibid.*, **10**, 549 (1957).
7. Bernstein, R. B., F. F. Cleveland, and F. L. Voelz. "Substituted Methanes. XVIII. Vibrational Spectra of $C^{13}H_3I$," *J. Chem. Phys.* **22**, 193 (1954).
8. Bethell, D., and V. Gold. "Aromatic Alkylation. Part I. The Kinetics of the Acid-catalysed Aralkylation by Diarylmethanols in Acetic Acid Solution," *J. Chem. Soc.*, **1958**, 1905.
9. Bigeleisen, J. "Isotope Effect in the Decarboxylation of Labelled Malonic Acids," *J. Chem. Phys.*, **17**, 425 (1949).
10. Bigeleisen, J. "The Relative Reaction Velocities of Isotopic Molecules," *J. Chem. Phys.*, **17**, 675 (1949).
11. Bigeleisen, J., and L. Friedman. "C^{13} Isotope Effect in the Decarboxylation of Malonic Acid," *J. Chem. Phys.*, **17**, 998 (1949).
12. Bigeleisen, J., and M. Goeppert Mayer. "Calculation of Equilibrium Constants for Isotopic Exchange Reactions," *J. Chem. Phys.*, **15**, 261 (1947).
13. Bigeleisen, J., and M. Wolfsberg. "Temperature Independent Factor in the Relative Rates of Isotopic Three Center Reactions (and Errata note)," *J. Chem. Phys.*, **21**, 1972 (1953); **22**, 1264 (1954).
14. Bigeleisen, J., and M. Wolfsberg. Chapter II. "Theoretical and Experimental Aspects of Isotope Effects in Chemical Kinetics," in I. Prigogine (ed.), *Advances in Chemical Physics*, Vol. I, Inter-

science Publishers, Inc., New York; Interscience Publishers, Ltd., London, 1958, p. 15.

15. Binks, J. H., and J. H. Ridd. "The Mechanism of the Coupling of Diazonium Salts with Heterocyclic Compounds. Part II. The Reaction of the Neutral Indole Molecule," *J. Chem. Soc.*, **1957**, 2398.

16. Bonner, L. G., and R. Hofstadter. "Vibration Spectra and Molecular Structure. IV. The Infra-Red Absorption Spectra of the Double and Single Molecules of Formic Acid," *J. Chem. Phys.*, **6**, 531 (1938).

17. Bonner, T. G., F. Bowyer, and G. Williams. "Nitration in Sulphuric Acid. Part IX. The Rates of Nitration of Nitrobenzene and Pentadeuteronitrobenzene," *J. Chem. Soc.*, **1953**, 2650.

18. Bonner, T. G., and J. M. Wilkins. "The Cyclodehydration of Anils. Part II. The Hydrogen Isotope Effect in the Cyclodehydration of 2-Anilinopent-2-en-4-one," *J. Chem. Soc.*, **1955**, 2358.

19. Brown, H. C., and G. A. Russell. "The Photochlorination of 2-Methylpropane-2-d and α-d_1-Toluene; the Question of Free Radical Rearrangement or Exchange in Substitution Reactions," *J. Am. Chem. Soc.*, **74**, 3995 (1952).

20. Brügel, W. *Einführung in die Ultrarotspektroskopie*, 2. Aufl., Verlag von Dr. Dietrich Steinkopff, Darmstadt, 1957.

21. Bryce-Smith, D., V. Gold, and D. P. N. Satchell. "The Hydrogen Isotope Effect in the Metallation of Benzene and Toluene," *J. Chem. Soc.*, **1954**, 2743.

22. Brynko, C., G. E. Dunn, H. Gilman, and G. S. Hammond. "Isotope Effect in the Hydrolysis of Triphenylsilane-d," *J. Am. Chem. Soc.*, **78**, 4909 (1956).

23. Buist, G. J., and M. L. Bender. "Carbon-14 Kinetic Isotope Effects. IV. The Effect of Activation Energy on Some Carbon-14 Kinetic Isotope Effects," *J. Am. Chem. Soc.*, **80**, 4308 (1958).

24. Bunnett, J. F., E. W. Garbisch, Jr., and K. M. Pruitt. "The 'Element Effect' as a Criterion of Mechanism in Activated Aromatic Nucleophilic Substitution Reactions," *J. Am. Chem. Soc.*, **79**, 385 (1957).

25. Bunton, C. A., D. P. Craig, and E. A. Halevi. "The Kinetics of Isotopic Exchange Reactions," *Trans. Faraday Soc.*, **51**, 196 (1955).

26. Convery, R. J., and C. C. Price. "Further Data on the Free Radical Phenylation of 2,4-Dinitrotritiobenzene," *J. Am. Chem. Soc.*, **80**, 4101 (1958).

27. De la Mare, P. B. D., T. M. Dunn, and J. T. Harvey. "The Kinetics and Mechanisms of Aromatic Halogen Substitution. Part IV. Rates of Bromination of Benzene and Hexadeuterobenzene by Aqueous Hypobromous Acid Containing Perchloric Acid," *J. Chem. Soc.*, **1957**, 923.

28. Denney, D. B., and P. P. Klemchuk. "Deuterium Isotope Effects in Some Acid-catalyzed Cyclizations of 2-Deuterio-2'-Carboxybiphenyl," *J. Am. Chem. Soc.*, **80**, 3285 (1958).

29. Deno, N. C., and R. W. Taft, Jr. "Concentrated Sulfuric Acid–Water," *J. Am. Chem. Soc.*, **76**, 244 (1954).

30. DeTar, D. F., and M. N. Turetzky. "The Mechanisms of Diazonium Salt Reactions. I. The Products of the Reactions of Benzenediazonium Salts with Methanol," *J. Am. Chem. Soc.*, **77**, 1745 (1955).

31. Dibeler, V. H., and R. B. Bernstein. "Isotope Effect on Dissociation Probabilities in the Mass Spectra of Chloroform and Chloroform-*d*," *J. Chem. Phys.*, **19**, 404 (1951).

32. Dole, M. *Introduction to Statistical Thermodynamics*, Prentice-Hall, Inc., Englewood Cliffs, N. J., 1954, Appendix, Table 6, p. 240.

33. Eliel, E. L., Z. Welvart, and S. H. Wilen. "Isotope Effects in the Free Radical Arylation of Aromatic Hydrocarbons," *J. Org. Chem.*, **23**, 1821 (1958).

34. Eliel, E. L., P. H. Wilken, F. T. Fang, and S. H. Wilen. "Reactions of Free Radicals with Aromatics. I. The Unimportance of the 'Identity Reaction' with Benzylic Radicals," *J. Am. Chem. Soc.*, **80**, 3303 (1958).

35. Eyring, H., and F. W. Cagle, Jr. "The Significance of Isotopic Reactions in Rate Theory," *J. Phys. Chem.*, **56**, 889 (1952).

36. Friedman, L., and J. Bigeleisen. "Oxygen and Nitrogen Isotope Effects in the Decomposition of Ammonium Nitrate," *J. Chem. Phys.*, **18**, 1325 (1950).

37. Frost, A. A., and R. G. Pearson. *Kinetics and Mechanism*, John Wiley & Sons, Inc., New York; Chapman & Hall, Ltd., London, 1953, (*a*) p. 178.

38. Gilman, H., G. E. Dunn, and G. S. Hammond. "An Unusual Isotope Effect," *J. Am. Chem. Soc.*, **73**, 4499 (1951).

39. Glasstone, S., K. J. Laidler, and H. Eyring. *The Theory of Rate Processes*, McGraw-Hill Book Company, Inc., New York and London, 1941, (*a*) p. 100, (*b*) footnote on p. 402.

40. Grimison, A., and J. H. Ridd. "The Significance of a Hydrogen-Isotope Effect in the Orientation of Electrophilic Substitution in Glyoxaline," *Proc. Chem. Soc.*, **1958**, 256.

41. Gronowitz, S., and K. Halvarson. "The Hydrogen Isotope Effect in the Metalation of Thiophene," *Arkiv Kemi*, **8**, 343 (1955).

42. Grovenstein, E., Jr., and D. C. Kilby. "Kinetic Isotope Effect in the Iodination of 2,4,6-Trideuterophenol," *J. Am. Chem. Soc.*, **79**, 2972 (1957).

43. Halevi, E. A. "Secondary Hydrogen Isotope Effects as a Criterion of Mechanism," *Tetrahedron*, **1**, 174 (1957).

44. Halvarson, K., and L. Melander. "Orientation and Hydrogen Isotope Effect in Nitration of Anisole," *Arkiv Kemi*, **11**, 77 (1957).

45. Hammond, G. S. "A Correlation of Reaction Rates," *J. Am. Chem. Soc.*, **77**, 334 (1955).

46. Harris, G. M. "Kinetics of Isotope Exchange Reactions," *Trans. Faraday Soc.*, **47**, 716 (1951).

47. Harrison, A. G., and H. G. Thode. "The Kinetic Isotope Effect in the Chemical Reduction of Sulphate," *Trans. Faraday Soc.*, **53**, 1648 (1957).

48. Herzberg, G. *Molecular Spectra and Molecular Structure. II. Infrared and Raman Spectra of Polyatomic Molecules*, D. Van Nostrand Company, Inc., Toronto, New York, London, 1945, (*a*) p. 195, Table 51.

49. Hine, J., and N. W. Burske. "The Kinetics of the Base-catalyzed Deuterium Exchange of Dichlorofluoromethane in Aqueous Solution," *J. Am. Chem. Soc.*, **78**, 3337 (1956); J. Hine, N. W. Burske,

M. Hine, and P. B. Langford. "The Relative Rates of Formation of Carbanions by Haloforms," *ibid.*, **79**, 1406 (1957).

50. Hirschfelder, J. O., and E. Wigner. "Some Quantum-Mechanical Considerations in the Theory of Reactions Involving an Activation Energy," *J. Chem. Phys.*, **7**, 616 (1939).

51. Hodnett, E. M., and J. J. Flynn, Jr. "A Study of the Decomposition of *p*-Nitrophenethyltrimethylammonium Iodide by Means of the Hydrogen Isotope Effect," *J. Am. Chem. Soc.*, **79**, 2300 (1957).

52. Hughes, E. D., C. K. Ingold, and R. I. Reed. "Kinetics and Mechanism of Aromatic Nitration. Part II. Nitration by the Nitronium Ion, NO_2^+, Derived from Nitric Acid," *J. Chem. Soc.*, **1950**, 2400.

53. Johnson, R. R., and E. S. Lewis. "Isotope Effect as a Measure of Transition-State Hindrance," *Proc. Chem. Soc.*, **1958**, 52.

54. Johnston, H. S. "Kinetic Isotope Effect Involving Methyl Radicals," *Science*, **128**, 1145 (1958).

55. Kaplan, L. "Hydrogen Isotope Effect in the Bromine Oxidation of Ethanol," *J. Am. Chem. Soc.*, **76**, 4645 (1954).

56. Kaplan, L. "Isotope Effects in the Chromic Acid Oxidation of 2-Propanol-2-*t*," *J. Am. Chem. Soc.*, **77**, 5469 (1955).

57. Kaplan, L. "Deuterium Isotope Effects in the Bromine Oxidation of Ethanol and of Acetaldehyde," *J. Am. Chem. Soc.*, **80**, 2639 (1958).

58. Kaplan, L., and K. E. Wilzbach. "Hydrogen Isotope Effect in the Hydrolysis of Triphenylsilane," *J. Am. Chem. Soc.*, **74**, 6152 (1952); "Hydrogen Isotope Effects in the Alkaline Cleavage of Triorganosilanes," **77**, 1297 (1955).

59. Lauer, W. M., and W. E. Noland. "The Nitration of Monodeuterobenzene," *J. Am. Chem. Soc.*, **75**, 3689 (1953).

60. Leo, A., and F. H. Westheimer. "The Chemistry of Diisopropyl Chromate," *J. Am.Chem. Soc.*, **74**, 4383 (1952).

61. Lewis, E. S. "Isotope Effects and Hyperconjugation," *Tetrahedron*, **5**, 143 (1959).

62. Lewis, E. S., and G. M. Coppinger. "The Nature of Participation of Hydrogen in Solvolytic Reactions," *J. Am. Chem. Soc.*, **76**, 4495 (1954).

63. Martin, D. C., and J. A. V. Butler. "The Dissociation Constants of Some Nitrophenols in Deuterium Oxide," *J. Chem. Soc.*, **1939**, 1366.

64. Melander, L. "On the Mechanism of Electrophilic Aromatic Substitution — An Investigation by Means of the Effect of Isotopic Mass on Reaction Velocity," *Arkiv Kemi*, **2**, 211 (1950).

65. Melander, L. "Nuclear Chemical Investigation of the Alcohol Reduction of Diazonium Salts," *Arkiv Kemi*, **3**, 525 (1951).

66. Melander, L. "Note on the Kinetics of Exchange of Isotopes with Appreciable Isotope Effect," *Arkiv Kemi*, **7**, 287 (1954).

67. Melander, L. C. S. "The Use of Nuclides in the Determination of Organic Reaction Mechanisms," *The Peter C. Reilly Lectures in Chemistry*, Vol. XI, University of Notre Dame Press, Notre Dame, Indiana, 1955.

68. Melander, L., and P. C. Myhre. "The Interpretation of the h_0 Correlation in Some Acid-catalyzed Reactions," *Arkiv Kemi*, **13**, 507 (1959).

69. Melander, L., and S. Olsson. "Kinetic Isotope Effect in Isotopic Exchange. Electrophilic Exchange of Hydrogen in Benzene and Toluene," *Acta Chem. Scand.*, **10**, 879 (1956).

70. Murray, A., III, and D. L. Williams. *Organic Syntheses with Isotopes*, Parts I and II, Interscience Publishers, Inc., New York; Interscience Publishers Ltd., London, 1958.

71. Olah, G. A., and S. J. Kuhn. "Aromatic Substitution. IV. Protonated and Deuterated Alkylbenzene Tetrafluoroborate Complexes," *J. Am. Chem. Soc.*, **80**, 6535 (1958); G. A. Olah, A. E. Pavlath, and J. A. Olah. "Aromatic Substitution. V. The Synthesis of a Protonated Toluene Tetrafluoroborate Complex," *ibid.*, **80**, 6540 (1958); G. A. Olah and S. J. Kuhn. "Aromatic Substitution. VI. Intermediate Complexes and the Reaction Mechanism of Friedel-Crafts Alkylations and Acylations," *ibid.*, **80**, 6541 (1958).

72. Olsson, S. "Acid-induced Aromatic Hydrogen Exchange. I. Determination of the Exchange Rates of Deuterium and Tritium Against Protium in Benzene and Toluene," *Arkiv Kemi*, **14**, 85 (1959), and following papers of the same series.

73. Pitzer, K. S. "Carbon Isotope Effect on Reaction Rates," *J. Chem. Phys.*, **17**, 1341 (1949).

74. Purlee, E. L. "On the Solvent Isotope Effect of Deuterium in Aqueous Acid Solutions," *J. Am. Chem. Soc.*, **81**, 263 (1959).

75. Redlich, O. "Eine allgemeine Beziehung zwischen den Schwingungsfrequenzen isotoper Molekeln," *Z. physik. Chem.*, **B28**, 371 (1935).

76. Reitz, O. "Über die Loslösung von Protonen und Deuteronen aus organischen Molekülen bei allgemeiner Basenkatalyse, untersucht an Hand der Bromierung des Nitromethans," *Z. physik. Chem.*, **A176**, 363 (1936).

77. Reitz, O. "Zur allgemeinen Säuren- und Basenkatalyse in leichtem und schwerem Wasser. Die durch Wasserstoffionen katalysierte Bromierung des Acetons," *Z. physik. Chem.*, **A179**, 119 (1937).

78. Reitz, O., and J. Kopp. "Zur Säure- und Basenkatalyse in leichtem und schwerem Wasser. III. Die Bromierung des Acetons, katalysiert durch undissoziierte Säuren und durch Acetationen," *Z. physik. Chem.*, **A184**, 429 (1939).

79. Roe, A., and M. Hellmann. "Determination of an Isotope Effect in the Decarboxylation of Malonic-1-C[14] Acid," *J. Chem. Phys.*, **19**, 660 (1951).

80. Roginsky, S. Z. *Theoretical Principles of Isotope Methods for Investigating Chemical Reactions*, Academy of Sciences USSR Press, Moscow, 1956, English translation by Consultants Bureau, Inc., New York, 1957, Chapter IV.

81. Ropp, G. A. "Isotope Evidence for the Mechanisms of Decarbonylation of Three Carboxylic Acids in Sulfuric Acid," *J. Am. Chem. Soc.*, **80**, 6691 (1958).

82. Ropp, G. A., and V. F. Raaen. "A Comparison of the Magnitudes of the Isotope Intermolecular Effects in the Decarboxylations of

Malonic-1-C^{14} Acid and Malonic-2-C^{14} Acid at 154°," *J. Am. Chem. Soc.*, **74**, 4992 (1952).

83. Ropp, G. A., A. J. Weinberger, and O. K. Neville. "The Determination of the Isotope Effect and Its Variation with Temperature in the Dehydration of Formic-C^{14} Acid," *J. Am. Chem. Soc.*, **73**, 5573 (1951).

84. Rule, C. K., and V. K. La Mer. "Dissociation Constants of Deutero Acids by E.m.f. Measurements," *J. Am. Chem. Soc.*, **60**, 1974 (1938).

85. Russell, G. A. "Solvent Effects in the Reactions of Free Radicals and Atoms. IV. Effect of Aromatic Solvents in Sulfuryl Chloride Chlorinations," *J. Am. Chem. Soc.*, **80**, 5002 (1958).

86. Schubert, W. M., and H. Burkett. "Acid-Base Catalysis in Concentrated Acid Solution. Deuterium Isotope Effects in the Decarbonylation of Aromatic Aldehydes," *J. Am. Chem. Soc.*, **78**, 64 (1956).

87. Schubert, W. M., and P. C. Myhre. "General *vs.* Specific Acid-Base Catalysis in Strong Mineral Acid Solution. Aromatic Decarbonylation," *J. Am. Chem. Soc.*, **80**, 1755 (1958).

88. Shiner, V. J., Jr. "Substitution and Elimination Rate Studies on Some Deutero-isopropyl Bromides," *J. Am. Chem. Soc.*, **74**, 5285 (1952).

89. Shiner, V. J., Jr. "Solvolysis Rates of Some Deuterated Tertiary Amyl Chlorides," *J. Am. Chem. Soc.*, **75**, 2925 (1953).

90. Shiner, V. J., Jr. "Deuterium Isotope Rate Effects and Steric Inhibition of Hyperconjugation," *J. Am. Chem. Soc.*, **78**, 2653 (1956).

91. Shiner, V. J., Jr. "Deuterium Isotope Effects and Hyperconjugation," *Tetrahedron*, **5**, 243 (1959).

92. Shiner, V. J, Jr., and C. J. Verbanic. "The Effects of Deuterium Substitution on the Rates of Organic Reactions. IV. Solvolysis of *p*-Deuteroalkyl Benzhydryl Chlorides," *J. Am. Chem. Soc.*, **79**, 373 (1957).

93. Streitwieser, A., Jr., R. H. Jagow, R. C. Fahey, and S. Suzuki. "Kinetic Isotope Effects in the Acetolyses of Deuterated Cyclopentyl Tosylates," *J. Am. Chem. Soc.*, **80**, 2326 (1958).

94. Swain, C. G., T. E. C. Knee, and A. J. Kresge. "Relative Reactivities of Toluene, Toluene-α,α,α-d_3 and Toluene-α-t in Electrophilic Nitration, Mercuration and Bromination," *J. Am. Chem. Soc.*, **79**, 505 (1957).

95. Swain, C. G., E. C. Stivers, J. F. Reuwer, Jr., and L. J. Schaad. "Use of Hydrogen Isotope Effects to Identify the Attacking Nucleophile in the Enolization of Ketones Catalyzed by Acetic Acid," *J. Am. Chem. Soc.*, **80**, 5885 (1958).

96. Urey, H. C. "The Thermodynamic Properties of Isotopic Substances," *J. Chem. Soc.*, **1947**, 562.

97. Walling, C. *Free Radicals in Solution*, John Wiley & Sons, Inc., New York; Chapman & Hall, Ltd., London, 1957.

98. Walling, C., and B. Miller. "The Relative Reactivities of Substituted Toluenes Toward Chlorine Atoms," *J. Am. Chem. Soc.*, **79**, 4181 (1957).

99. Westheimer, F. H., and N. Nicolaides. "The Kinetics of the Oxidation of 2-Deuteropropanol-2 by Chromic Acid," *J. Am. Chem. Soc.*, **71**, 25 (1949).

100. Wheland, G. W. *Resonance in Organic Chemistry*, John Wiley & Sons, Inc., New York; Chapman & Hall, Ltd., London, 1955, Appendix, p. 695.

101. Wiberg, K. B. "The Deuterium Isotope Effect," *Chem. Revs.*, **55**, 713 (1955).

102. Wiberg, K. B., and L. H. Slaugh. "The Deuterium Isotope Effect in the Side Chain Halogenation of Toluene," *J. Am. Chem. Soc.*, **80**, 3033 (1958).

103. Wilen, S. H., and E. L. Eliel. "Reactions of Free Radicals with Aromatics. II. Involvement of Ring Hydrogens in the Reaction of Methyl Radicals with Alkylbenzenes," *J. Am. Chem. Soc.*, **80**, 3309 (1958).

104. Wood, R. W., and D. H. Rank. "The Raman Spectrum of Heavy Chloroform," *Phys. Rev.*, **48**, 63 (1935).

105. Zollinger, H. "Kinetische Wasserstoffisotopeneffekte und allgemeine Basenkatalyse der Azokupplung"; "Abhängigkeit des kinetischen Isotopeneffektes der Azokupplung von Basenkonzentration und Diazokomponente"; "Über die Natur der Protonabspaltung bei Azokupplungen," *Helv. Chim. Acta*, **38**, 1597, 1617, 1623 (1955).

106. Zollinger, H. "The Mechanism of Electrophilic Aromatic Substitution," *Experientia*, **12**, 165 (1956).

Index

Absolute reaction rate, 9 ff.
  limitation of theory of, 7
Acceptor, for hydrogen, 24 ff., 32, 67 ff.
Acetate ion, as hydrogen acceptor, 77, 78
Acetic acid, as proton donor, 78
Acetolysis, of cyclopentyl toluene-p-sulfonate, 90 ff.
  of methyl-p-tolylcarbinyl chloride, 94
Acetone, bromination of, 77
Acid-base catalysis, 125 ff.
Activated complex; *see* Transition state
Activation energy, 18, 44, 45, 75, 123, 145
  intermediate-temperature difference in experimental, 45
  low-temperature difference in experimental, 44
Activation equilibrium, 10, 12 ff., 41, 45, 142, 163
  equilibrium constant of, 13 ff., 34
Activity coefficient, 10, 15, 41 ff.
Alcohols, oxidation of, 100 ff.
Alkaline cleavage, of triphenylsilane, 80 ff.
  of tri-n-propylsilane, 83 ff.
Alkylation, of anisole, 111
p-Alkylbenzhydryl chlorides, solvolysis of, 95
Ammonium nitrate, decomposition of, 158 ff.
Amount of reaction, 49 ff.
Anharmonicity, vibrational, 9 n., 96

2-Anilinopent-2-en-4-one, cyclodehydration of, 116
Anisole, alkylation of, 111
  nitration of, 110
Aromatic substitution, electrophilic, 95, 99, 107 ff.
  metalation, 99 ff.
  nucleophilic, 122 ff.
  radical, 123 ff.
Arylation, of m-dinitrobenzene, 125
Azo coupling, of aromatic compounds, 113 ff., 115 ff.

Bending frequencies, approximated by zero, 21 ff., 35 ff., 67, 75
Bending vibrations, of bond being ruptured, 19, 21 ff., 35 ff.
  of three-center transition state, 30
Benzene, bromination of, 111
  electrophilic hydrogen exchange in, 118 ff.
  metalation of, 99 ff.
  nitration of, 110
Bigeleisen-Goeppert Mayer heavy-atom approximation, 36 ff., 130 ff., 165
Bigeleisen-Wolfsberg temperature-independent factor, 39, 146, 151
Bigeleisen's treatment, 32 ff., 128, 139 ff., 146 ff., 151, 152, 161 ff., 165
Branching ratio, 54, 73
Bromination, of acetone, 77
  of benzene, 111
  of 2-naphthol-6,8-disulfonic acid, 114

177